MIND FROM MATTER:

A Short History of Humanity

Kenneth F. Deboer

Mind From Matter: A Short History of Humanity

Kenneth F. Deboer

copyright 2018 Kenneth F. Deboer

ISBN-13: 9781505455151
ISBN-10: 1505455154

Printed by CreateSpace

Cover photographs by the author

CONTENTS

CONTENTS (Continued)

ACKNOWLEDGEMENTS

This book was instigated by and is dedicated, with love and gratitude, to my grandson Brian Clark. Himself a writer, he provided the impetus and intellectual underpinnings and provided advice and incisive editing suggestions throughout. It was made immeasurably better by his strenuous efforts, although I bear all responsibility for the end result. He did this even though he is in substantial disagreement with many portions of the philosophy. Thank you, Brian. I also thank another grandson, Mark Deboer, who provided great help in developing the Cover and also the illustrations. He made this job possible and fun. (BTW, he is still in public school.) I am grateful too to Dr. Ralph Johnson of Gunnison, Colorado, who kindly read the manuscript. He contributed innumerable suggestions for improvement, which spared the reader from many a painful passage. Additionally, I owe a debt to a large number of sources from a variety of arenas that I am unable to acknowledge because I can't remember who or where a particular tidbit came from. Some of the ideas, sayings, thoughts and facts that I use in passing throughout the book come either from way back somewhere or in an online or journal or other piece that bears reporting, but I can't now identify exactly. I would appreciate corrective feedback from those who might legitimately desire credit but that I failed to identify. In fact, I apologize in advance for leaving out references to so many excellent scientific, technical, economic and social and other workers who are at this very moment publishing on and blazing so many interesting new trails.

PREFACE

This is admittedly a funny kind of book. One not meant for the specialist, but Everyman, any kind of person with even a passing interest in the nature and history of man and all his previous activities. It is my own personal take and overview of the most notable events that occurred in the world since the beginning. The book encompasses 'everything'; but the obvious focus is on The Most Notable event to ever occur on this planet—Man and his Mind. The look-in is from 10,000 feet and the travel is fast and out of necessity, quite superficial. The aim is to provide mainly a simple, selective 'movie', from my personal perspective, of the activities on earth, principally from the beginning of Man's footprint down to today.

I start with the beginnings of life and trace the evolution of life on earth, especially the origins and nature of the Mind within Man, with its quirks and limits. His invention of civilizations is briefly traced to illustrate the workings out of his nature in the recent past (our only, singular, past). Then we arrive at Modern Man, the dictator of the world, as he hogs now nearly 40 percent of all the earth's photosynthetic output either directly or indirectly for his use. He did it by specializing in the Mind. Finally, I try to imagine where he is heading and suggest some general way by which that future might be improved. I make no claim to be scientifically pure or comprehensive but want mainly to illustrate, with some limited more or less representative samples of scientific data or theory to indicate the gist of where most contemporaries seem to 'be at' overall with respect to our technology. Most of all the aim is to give the general reader context, background and historical underpinnings of the more or less standard general understanding of the status of human nature in the present world. The hope is that it might encourage further and deeper exploration and give a few insights and a little pleasure along the way.   K. Deboer, Livingston, Montana 2018

CHAPTER I.
MAN: WHO, WHAT, WHERE, HOW?

Introduction

These may not be the best of times, but for many people these seem to be the worst of times. Anxiety! Lonely! Rat race! How does one stake out a secure and happy life in the crowded jangle of modern life? Older generations cower under the pelting bombardment of phantasmogoric technology and lightning speed of change. A goodly fraction of eight billion people on earth wage daily internal and/or external battles against poverty, war, upheavals, privation, keeping their head above water, anonymity, hopelessness, or just against society itself, or demons within. The problems in the world seem immense and intense, the directions muddled and uncertain; as E. O. Wilson put it "We have created a star wars civilization with Stone Age emotions, medieval institutions and godlike technology."

We 're all familiar with this modern – and eternal – litany. But why so intense now? Given that our physical surroundings are brighter than ever, at least from the standpoint of 500 years ago? What is the cause of it? Is there any remedy? "Who do we think we are?" Throughout the ages many have attempted to provide their own answers through science, religion, Oneness, inner searching, art, expansive social contracts, etc. Fair enough: it is useful to have some tools to work with to progress.

What does not seem tenable forever is to keep wearing deeper the same ruts without at least seriously attempting to eventually make things even better – i.e., progress. Survival of high civilizations, or even humans, is not inevitable. We ask, can science have anything useful to say about Man, his psyche and mentality? Or on the other hand, is it helpful if some scientists or 'realists' deny that any fundamentally true and progressive and usable truths can come from psyche, art or philosophy. At the very least we may say these are confusing and dangerous times.

It is the purpose of this work to try to frame these kinds of questions in more or less historical, that is evolutionary, contexts. To understand what and where we are, it will help if we better understand what once was, what is now, and what will be. I will be asking ' where did we come from, what are we made of and what is our nature'. My training in biology colors my narrative and bids me state upfront a couple of central facts and tenets. First, we must agree that we arose from, and are and will continue to be part of Nature. Also, we should always bear in mind that (a) Man – including his astounding Mind – evolved by natural selection from tribal African ancestors a million or so years ago and is genetically largely unchanged since; and (b) the incredible cultural evolution of his Mind over the past 50,000 years or so is now his most overwhelming feature. Man alone on this planet, has this central and overwhelming feature of Mind and a special kind of Consciousness.

Consciousness itself, of course, arose even earlier than us; in fact much earlier and we are not alone in possessing some degree of that mysterious faculty. I am confident though, that exceptionally high degrees of self- awareness, intentionality, high rational predictive and abstract thought belong to our species and ours alone.

Importantly, these traits arose and were useful only because it aided survival and reproduction. That is also why, since we still have and operate that very same ancient brain, our conscious thoughts are therefore still mostly driven by ancient emotions and deep rooted desires; hence much of our angst and anguish. Our brains, i.e. our nervous tissue, wherein lies our emotional machinery, intellect and consciousness, are maybe a million years old and they have not appreciably changed over the last 50,000 years or so. However, our *minds* have expanded (evolved) a millionfold due to the accretions and demands of our society and our culture.

While we will ultimately be concerned mainly with our high mental powers, I will first sketch our physical evolution, including all life on earth. Thus, we'll first take a quick whirlwind tour of the earth, including man and his whole history, including his accomplishments, thoughts and mind. Then, in following chapters we will further explore our conscious mental abilities which arose in our ancestors long before much of its potential could possibly be used for anything beyond simple living and the struggle for survival. It was not designed for self- examination, or mathematics, or art, or religion or any of the activities that now constitute almost the whole point of humanity. It was a lonely, unrefereed, often wrenching struggle of the ape-men over thousands of generations. And now—here we are!

Genesis

At the outset of our vast history review, we see from the retrospective of the educated man today that three stupendous, totally unpredictable events happened; all else is pretty much established, prosaic carrying on of Nature; (1) " where did the universe come from in the first place?" (because in a readily understandable way all else flows from that); (2) The second astounding event was the emergence of the organized matter called life, especially the first chemical blob that was definitely alive and from which all other forms of life developed naturally; (not automatically or inevitably, or evenly but still, naturally); (3) The third mysterious event is the newest and biggest one; the self-aware Mind.

Although an invisible form of the biosphere, our Mind and the totally flabbergasting phenomenon of modern man and his works is the first and foremost Wonder of this world. By this invention, in a twinkling of an eye, one species has totally taken over the whole planet and now, if a recognizable biosphere is to continue, must needs take control of the whole earth and everything in it. Man, the God species. We will come back later to consider the critical 'bright line' (actually a murky labyrinthine pathway) that separates us from all previous animals.

Let's begin with a brief review of the universe itself and our little space in it. The Universe is currently thought to have begun about 14 billion years ago with a Big Bang. Back then there existed really no thing; nothing we could recognize, no rocks, no water, no life, no molecules or even atoms. Maybe just highly concentrated pure swirling energy or nascent quark soup, monopoles and other strange things.

However, EVERYTHING eventually came out of this 'nothingness', i.e. all the natural elements, energy and matter ('regular' and 'dark'), stars, planets, rocks, people; everything was made out of this initial raw material. This was the real SOURCE. As the British author Terry Pratchett explained "In the beginning there was nothing – – then it exploded!"

(Parenthetically, I will note an important corollary of the above facts. That is, this stuff is still IT. All that is, or will be, and space itself is bounded by the limits of these types of 'things' and their complicated interactions. Interestingly, it turns out too that the Universe, broadly speaking, is quite uniform everywhere. And, as Carl Sagan conclusively taught us, we ourselves are made of this primordial starstuff. Why even this stuff existed is unanswerable at present.

In any event, just 4.6 billion years ago, our own little particular part of the universe formed. Our young solar system with our sun and her planets, moons, satellites, etc. came about from gravitational collapse of a huge mass of stellar gas and 'dust'. This fiery interstellar disc with our central star the sun, still fiery after all these years, rather soon chinked off various small orbiting balls of dust, rock, debris and 'fire' – – the planets. On the cooling earth, atoms of hydrogen, oxygen, gritty particles of sodium, silicon and the other elements combined and re-combined in myriads of ways. Water eventually formed in pools and lakes. What a world of difference from the present lush, green, hospitable land – – our own little lonely, lovely blue planet. And the main difference is the arising of the amazing thing called 4

4

Life. Life became possible due to the cooling, mineralization and the other familiar global geologic processes which have been occurring ever since. The stabilization of the solar orbits lead to Goldilocks days, nights and seasons, enabling the flourishing of life as we know it today.

Origins of the 'simple' life

How exactly Life arose is not known, or maybe not completely knowable scientifically, but the general features and timeline are very clear. We will pass through the story of evolution on the earth very quickly, as there is nothing really mystifying about any of it until we come to about a million years ago – the origin of the Mind of man.

As we all know, our bodies and all living things are composed of mostly molecular organic chemicals. Most of these are unique to life and are very highly organized in structure. The simplest and first of these more highly organized molecules arose most likely in water nearly 4 billion years ago. Chemical activity of the same type as continues today in, under and over the earth occurred then too, especially in some of the favorable pools. There, new collections of atoms and molecules accumulated to some degree. Carbon, hydrogen, oxygen and nitrogen were especially abundant at the surface, and since carbon and hydrogen are particularly chemically active, early on they formed larger combinations of molecules. Molecules, like methane, ammonia, carbon dioxide and water became abundant. From these comes quite naturally amino and also nucleic acids and other small organic molecules which are the critical building blocks of all living things today. Life itself, at its core is actually largely based on this constellation of these simple but amazing compounds. These compounds already came with the unique, intrinsic ability to form new types of chemicals, including polymers. Polymers are one of life's secrets and one of its biggest inventions. (There are

5

not many polymers occurring elsewhere in nature; small polymers are found, for example, in obsidian and some other rock types.)

What the actual form of the first 'life' on earth was is not clearly known, other than that "it" must have been a more or less large – albeit microscopic – aggregation of amino acid-like, ribonucleotide and other smallish molecules and their congeners or polymers. The key factor, as Dennett (2018) and also Gazzaniga (2018), Pattee (2016) and a few others have recently emphasized, is that they had a chemical (DNA) that could faithfully reproduce itself. Not only that, they had already the sine qua non of living things; that is, these molecules already contained a code (Information), which is a magical immaterial thing, a mere 'symbol', or stand-in for something else. It contains, completely blindly to itself, a 'rule' for specifying the actions of other molecules within the organism. By using this 'code' (which we now know consists of the 'information' of three consecutive nucleotides) they are able to produce other molecules which are the materials of their own 'body', i.e. amino acids, proteins, enzymes, etc. These are the carpenters which build all living things. This coding mechanism (in other words, Information, which is an immaterial thing) is what many people consider the secret to life, and the feature that sets life apart from all other things in the universe.

These core molecules, then, had the seemingly magical facility of being able to not only reproduce, but also persistently self- organize, carry out metabolism, and all other of life's characteristic activities. This original organism, by its very nature over countless millenia, carries out also (blind and unaided) R & D. In conjunction with the selective features of their environment over incredible periods of time, to ever increasing complexity and eventually to the miracle of all present life on earth.

These first biological things, were possibly something like a primitive virus today or, more likely an Archeon or bacteria. Even these molecular blobs have all the basic properties of life today. While they don't move, they do maintain their integrity within their environment by replacing or repairing any molecular disorganization (i.e. homeostasis) and eventually reproduce, even if a crude fashion. In a way, they are like 'parasites', which can 'thrive' and eventually reproduce under specific conditions, but still they displayed all the hallmarks and essentials of life.

In any event, these molecular aggregates occurred more than 3.7 billion years ago, which was only a billion years after the appearance of solid Earth. This was also a billion years before the present planetary orbits and magnetic fields had even become fixed. Yet complex chemicals had evolved and themselves had already formed self- replicating molecular aggregations. It then took 'only' a few hundred million years for these to further evolve into growing, self-reproducing, very complicated supra molecular blobs of Protoplasm; the very first appearance of such a substance, possibly anywhere. Remarkably, these blobs occurred already in a form recognizable today. These primitive corpuscular blobs, but still unmistakably organisms, were the first truly living things. These primitive 'cells', of course, were the ultimate ancestor of all current life too. The most amazing thing about them is that we actually can find some fossil remains of these earliest 'life blobs' today, so they are not completely imaginary.

Further evolution of these earliest proto-cellular life forms proceeded slowly under the very difficult environment of the early earth. It took 2 billion more years to advance to true modern cellular life, like the one- celled organisms similar to our present day protozoa. Even here great advances in structure and organization had taken place, as now these organisms were true cells ('Eukaryotes'), with sophisticated intracellular organelles, membranes and a nucleus. This type of cell is still, basically unchanged, *the* core unit in all forms of higher life.

A little later some of these cells 'invented' a means of using light, e.g., photons. These photosynthetic cells, the

7

progenitors of future plants have been discovered back to 6 billion years ago (BYA), and their effluent – oxygen – continued to build in the atmosphere. At 1.5 BYA these fancy new types of cells had advanced in complexity such that sex had also been invented. That is, some of these more complex first eukaryotic (modern) photosynthetic organisms could bud off parts of themselves – i.e. produce minute gametes (spores). These can conjoin to produce a new, additional "plant".

By 1000 million years ago (MYA), the first multicellular organisms had also developed. At first, this new type of organism was basically just an aggregation of many identical individual cells. It was still relatively motionless but nevertheless a complex organism made of separate, equal cellular units. It was similar, and of course ancestral, to all of the simple organisms of today, such as some slime molds or algae.

Artificial Life (?)

It is of course a long standing problem to account for the passage, which we have just cursorily traced, from consensually non-living to consensually living substance. The physical and chemical differences are extremely small but still immeasurable in its totality. While it is dubbed an Emergent property (a quite murky term at best), biologists of all stripes do not have any conceptual difficulty in ascribing a purely natural transition from non-living matter to living matter. The process is readily described as matter attaining a higher and higher level of organization, each step a small, natural one, aided by selection which kept the 'good' forms. A good deal of scientific thought and experiment has been leveled at this problem, beginning at the middle of the last century.

These types of experiments, many ongoing, show that the chemicals that are known to have been formed 3.8 billion years ago can be easily and naturally assembled into amino acids, sugars, phosphates and a goodly variety of other organic and inorganic molecules. By bringing these compounds tgether in a flask, under certain conditions they can spontaneously form still larger polymeric – chained – molecules, such as ribonucleic acids, peptides, polysaccharides and primitive membranous structures which mostly make up the structure of cells today.

Cyril Ponnamperuma for example, at the University of Maryland says "all our experiments show that under the right conditions the inorganic and organic worlds are inseparable. The ease with which the basic chemicals of life are made is incredible." Ponnamperuma as well as Sidney Fox at Miami U also found that in experiments like these, small peptide polymers are produced. Remarkably, these self-assemble themselves into tiny spheres that look a bit like modern microorganisms – and even more like the fossil protoplasmic blebs. Furthermore, under certain, even rather mild conditions these molecules can split into two 'daughters', powered only by the energy of mild heat. It seems most likely that the first primitive live microorganisms arose very much along these lines. Indeed, the concept of self-organization is a key concept which underlies all of life. As noted earlier, the molecules comprising Life, however, have that extra special 'trick' that enables more facile and rapid self-organization that gives us the astounding biota today, including the amazing Man. The shuffling of trillions of atoms and molecules which went on from the beginning--- and continues today—led little by little to increasingly more complex self-reorganizations which facilitated continuance (i.e. survival). This is how all change occurred in the living world. This is what is meant by evolution.

Some will insist that life cannot originate through nature alone but requires a supreme Creator or God or other supernatural phenomena. I would caution, though, that the prospects are very good of discovering (somewhat?) similar life elsewhere in the Universe. (In fact, the Mars Rover very

recently discovered several rather large organic molecules –
"biosignature molecules" –, such as long chain hydrocarbons
eerily like the fatty acids that are found in our own cell
membranes). Indeed, the prospects of life created in an
earthly lab in the not distant future seems almost certain.

Recently Markus Covert at Stanford reported the
simulation on a supercomputer of most of life's activity –
metabolism – going on within a bacterium, *Mycoplasma
genitalium*. Much more of this is on the way – – soon. Jack
Szostak, for another example, at Harvard has devoted his
whole lab to the artificial life effort and is reporting
substantial progress in assembling, from scratch, a living
cell. Indeed, based on scientific advances being made
already, there are very few biologists today who do not
believe that a living cell will soon be made synthetically.
Maybe of the rather primitive variety, but living
nevertheless, with metabolism, homeostasis, reproduction
and environmental interactions, which are the sine qua non
of life. This new Synthetic Biology will be, and is already,
quite a powerful enterprise. We will come back later to this
game changing kind of scientific progress.

In summary then, on the early earth we had aggregations of
self- assembling, self-reorganizing chemicals with
increasingly complex organization and reactions occurring.
It is conceived that these may have occurred first in the
plethora of hot chemical pools which contained a chemical
soup of the early molecules, including the organic ones as
noted earlier, like amino acids, etc. The interaction between
molecules therefore, morphed eventually, under the local
selective conditions, into highly organized self- replicating
units. As long as the 'food' and an energy source was still
available they could keep on maintaining themselves,
replicating and over time become more and more different –
i.e evolve – into more complex descendants. This is what is
expected from standard chemistry.

A common error so many people make in their thinking of
either chemistry or evolution, physical or biologic, is that it

is a totally random process, where chance is the only driving factor. However, chemical processes are governed largely by neither chance nor randomness. The nature of particles, atoms and molecules from the big bang on, is to react with others according to their intrinsic makeup and in accordance with well- known rules of physics. As long as some form of energy is added, chemicals react with each other under very simple, repeatable and easily discovered conditions to form more complex chemicals.

A simple illustration is when you toss a lump of salt – sodium chloride – into your pot of water you will always see the same reaction. The salt disappears and if you dig deeper you will see that individual atoms of electrically charged sodium and chloride are attached to groups of the water molecules. If one now adds another common chemical to that, (say, a dash of baking powder, or food coloring) you will get other different chemical reactions and also a totally different, but predictable, bunch of new, organized chemicals in your cup. These will be made molecule by molecule as the substituents happen, randomly and by chance, to bump into each other. However, the nature of the product will be determined solely by the nature of the 'bumping' types of material, as well as the environment, e.g. propinquity, temperature, pressure and so on. While chance is a big part of evolution at all levels, it is only one of the factors. It, plus enormous periods of time, and always-changing conditions (Nature's R & D) and selection pressure, leads (blindly) to change (i.e. evolution).

There will be no end to these automatic chemical and physical reactions as long as conditions allow, and not running out of reactable elements. So, evolution is an additive, R & D process, which can keep certain configurations, while it endlessly, naturally, tries out other – not forbidden – possibilities. When molecules arise, step by step that are more stable than others and if they interact with others of this or other simpatico type, then by natural selection that new structure will tend to persist in that

11

milieu. Same goes for all, including biological, evolution. Small advantageous characteristics are kept, the others disappear. This type of elemental structure of Nature appears to give a single direction to growth and change. Chance clearly plays a role as to which molecules happen to be near enough to react with each other, but the reactions are not random. It may truthfully be told that new molecules can arise by descent, i.e. evolve. Time, lots and lots and lots of it, and constant natural 'experimental' molecular reshuffling are the key factors.

The Rise of plants and animals on Earth

To return then to tracing evolution on the planet, by 570 MYA the super- continents were forming and re-forming. Significantly the orbit of the earth had by then more or less stabilized, creating a fixed daylight at 21 hours. This early period of time we're describing now is called the Cambrian Period, which was most marked by the 'rapid' evolution of all kinds of complex multicellular life in the oceans. First there were the soft bodied multicellular organisms, like the precursors of the jellyfish and similar organisms of today. (It may come as a shock to some, but the fact is that jellyfish already had by this time almost the same number of genes as man. And most of them are also of the same type, i.e., we share a majority of DNA. They are clearly kin, as are of course, all of the ensuing species.)

Most of the other marine invertebrate groups also sequentially made their first appearance during this period, like molluscs and similar. In fact, 'just' 100 million or so years passed before the first vertebrates, primitive fish, had already evolved in the various oceans. Shortly thereafter the first land organisms – fungi and small plants – arose. It is retrospectively obvious that evolution of living things was speeding up dramatically and in fact, this phenomenon has been speeding up ever since. Diversity feeds on diversity; and complexity feeds on complexity. (This goes double when we start considering the evolution of the Mind, especially cultural evolution.)

In the time span 443 MYA to 355 MYA the story of the earth continues to be the story of rapid proliferation of the number, types, complexity and size of living organisms. Coral reefs, a multitude of land plants, fish with jaws, sharks and a host of small land invertebrates, including insects, centipedes and others evolved from their immediate ancestors during this time. The first big plants also came early in this period, notably the ferns and gymnosperms. Huge forests of them became extensive around the world. Gradually, new and larger land animals appeared, most notably by 375 MYA the first 'fish' with legs. From them, shortly, came the first amphibians and then not too much later, lizards. This whole time period was marked by great growth, variation and spread of life. All of this new life occurred despite several major extinctions caused by meteors, major ice ages and the like. Life is aggressive, persistent and inventive. (Jane Hamilton, in " A Map of the World" puckishly asserted that the direction this ferocious life force took was by 'the guiding hand of chance'.)

Roughly for the next 100 million years after that (encompassing the Mississippian, Pennsylvanian, and Permian Periods for those keeping track) there was again rapid increases in number, size, complexity and diversity of life forms. Most of these newly arrived types were even rather familiar to us. Significant biological advances were seen in the rise of larger numbers and varieties of new species in all the existing groups. One major 'invention' was the *amniote* egg, arising first in the smaller land vertebrates. This advance meant that eggs need not be deposited in water to be fertilized externally but the eggs could be deposited directly for development on dry land. Most all vertebrates since that first one have these amniotic eggs, which also enables internal fertilization. The first winged insects appeared in this time too, some of them quite large. But perhaps most notable was the rise of good sized reptiles. Not

13

insignificantly, worldwide coal deposits were laid down too.

Late in this time frame, around 260 MYA, there was a changeable and apocalyptic period, marked by wild climate changes and also cataclysmic geologic disasters. Toward the end, the earth's temperature rose to its highest levels ever, due to tremendous volcanic activity on both land and sea spewing enormous amounts of long lasting carbon dioxide, methane and water vapor into the air. Oxygen levels fell from 30% to 12% and the Period ended with a devastating mass extinction (at 251 MYA). The final blow was delivered by a massive meteor, eliminating 70% of land animals.

Thus ended the Paleolithic Era, and began the Mesozoic Era, often called the era of the dinosaurs. This Era lasted until a mere 65 million years ago and is divided into three Periods. The first Period (Triassic) ran from 245 MYA to 208 MYA. Early on in the Triassic, the Pangea supercontinent started to break up and the survivors of the prior mass extinction, most notably small reptiles, rapidly repopulated and radiated. Small dinosaurs on land and giant marine ichthyosaurs and plesiosaurs in the sea flourished. Most notably, the first proto-mammals also appeared early on and by the end of the Period the first small mammal had arisen.

The next Period, the Jurassic, runs from 200 MYA to 145 MYA and is most famous as the golden age of the dinosaurs. The earth was unusually warm but not hot and plants like cycads, ginkgoes and conifers dominated the landscape. The large familiar dinosaurs ruled the land and the flying dinosaurs made their appearance too. The first birds appeared. Near the end of this Period, a most significant group appeared, small and inconspicuous at first, --true placental mammals.

The Cretaceous Period followed, from 145 to 65 MYA. This too was a warm period over most of the globe. The angiosperms (flowering plants) arose and thrived as did some of the dinosaurs still, including *Tyrannosaurus Rex*. Some of the present continents and land masses as well as

14

the Atlantic ocean appeared as the ice caps melted. This whole Period was racked by interstellar bombardments and at 65.5 MYA a huge meteorite crashed into the Yucatan peninsula. This resulted in global extinction of 85% of the land animals. This meteorite most notably knocked off the dinosaurs entire. This also marked the end of the Mesozoic Era and the beginning of the Cenozoic, which is our present Era.

Much of the history of the Cenozoic Era, from 65 MYA on is rather familiar in general outline to most people. It takes us from the dinosaur age to the present configuration of land masses as well as the modern animal and plant biota. This Era was no different in principle from the previous 3.5 billion years during which the earth experienced periodic physical poundings from both extraterrestrial and terrestrial natural forces. Many new Orders of life arose and some forms extinguished under the force of the pounding and competition. All this biological evolution resulted in the modern type of placental mammals, like rhinoceros, camels, early horses, elephants and long legged grazing animals. The first modern grasses arose and flourished. The Era also saw the origin and flourishing of the social insects along with great increases in variety and number of the flowering plants as well. By the end, the flora and fauna are modern, or at least the types would be recognizable to us.

Geologically, the Rocky Mountains were formed, Australia separated from Antarctica and the Antarctic polar ice caps formed. The Himalayas rose as the Indian subcontinent ran into Asia. Again, several large meteorites hit the earth at various places and times, creating havoc for a time and causing hundreds and sometimes thousands of species to disappear.

This then brings us down to the dawn of Man himself. Up to this point on our brief travel through time on Earth, from its very beginning 4.6 BYA and down to the edge of man, we have seen many fascinating and truly mind- boggling

15

events from beginning to end. In this survey I have introduced no new concepts or data but have just summarized the contents of Freshman Biology courses, Geology 101 and other sandard sources. The overwhelming main conclusion to draw from this panorama is that the earth, through ever more rapidly developing, often violent and disorderly progression (i.e. Nature's R & D), went from a lonely hot ball of rock, dust and debris to a lovely cool, green abode. Most of all, it writhes with the film of life; life, sprung from the earth itself and now spread out inch by inch, without interruption, across the whole globe. It moves and has 'character.' Life is its character. And it 'ends' with the lives of that most interesting character- Man.

The Planet and the Apes: On the Road To Man

So now we need to take a closer look at the time just before and after man's origin. We begin about 23.3 million years ago, in the Miocene Epoch. Here is where we get our first look at our own direct lineage – the apes. What should we look for when we examine the evolution of our own ancestors? There are two general categories to look at: 1) the physical changes in the apes which lead to man, and 2) the artifacts and environment which will indicate the crossing of the line from 'simple' animal mind to man's. In brief, some of the main physical changes which distinguish apes from man are (a) big brain, – most significant of course, (b) bipedal, (c) flattening of the face, (d) shortening of arms and lengthened legs, (e) flattened feet with short toes, (f) less massive bones, (g) hand and fingers more nimble, (h) loss of simian shelf and a host of other smaller changes. All of these are reflected in the fossil record. The other criteria, #2 above, which is much more difficult to establish by fossils and other indirect evidence, is the all-important mental transition from simpler minds and consciousness to self-

16

awareness and all the other high mental characteristics of Man.

The Miocene epoch was the golden age of the primates, with an abundance of species in Africa and especially, Eurasia. Early in the epoch the first group of monkeys appear and soon the higher primates, the *hominid*s. This latter group includes all the apes – and Man too, after he arises. During this period of time also, a mere 5 million or so years ago, North and South America became joined by the isthmus, and animals and plants began to enter South America. Significantly, the middle and late Miocene, and continuing on into the next epoch (the Pliocene) saw a cooling and drier period with substantial deforestation. Grasslands and savannahs expanded around the world, creating them as they pretty much are today Thus, in parts of Africa considerable environmental pressure and selection of the early primates culminated in several hominid species having developed special walking habits. (All species of this type however, were knuckle walkers, with limited upright locomotion. All of the current primates, except man, are still either partly or wholly arboreal and/or knuckle walkers.)

These advanced primates had a bit larger brain, lacked a tail and were clearly on the line to man. 'Taillessness' was actually a critical feature for later evolutionary advances as it helped enable true upright walking. Thus, some of the direct predecessor groups leading to man were already largely bipedal and also their hands were crudely adapted for grasping. By about 5.8 MYA at least three related groups were alive and floridly exhibiting these transition characteristics between the hominids and the *hominins* (Man alone). These three intriguing hominid fossil groups are named *Orrorin, Sahelanthropus* and *Ardipithecus,* each of which have proponents claiming them to be the first on the path leading to the first *hominins*. At any rate one of these most likely gave rise to the branch from which man directly derives. One such branch descending from this above group was the pivotal group called *"Australopithecus"*. This

17

important group is considered by many to actually be the first ancestral man. This group was the first to have entirely left the trees and to have truly walked on two legs. The original environmental impetus was the changing climate in Africa and the diminishing of forest areas. But why is upright walking evolutionarily advantageous?

There are many theories. A long standing one is that it freed the hand to make, carry and use tools. This is an especially useful feature for the prehuman ancestors like the ones noted above in that it frees the hands to also carry food. If one has to forage rather far and wide to gather food and then carry it to a safe place, one could carry more in the hands than in the mouth, which is the case with virtually all other advanced animals. In addition, traveling on two feet enables a high running speed and longer range. In fact, humans are the world's greatest long distance and endurance runners. Few large mammals can run six hours straight, or 25 miles at a time. This is no mean advantage; a human can run down hardly any animal directly, but by doggedly pursuing it over a long time, it can eventually bring to heel almost any animal. Especially if cooperative hunting is employed, which it was fairly early on. It is likely that this athletic prowess was used often to secure food and thus, eventually, enable *Us*.

It is also significant that early man at this point was one of only a handful of animals that routinely mixes the two means of obtaining and consuming food, i.e. hunting <u>and</u> gathering. Food, of course represents the main necessity – and threat – of survival. Gathering of fruits, nuts, berries, grasses and vegetables still was the main source of food for the scattered tribes of ape-men, but it was routinely supplemented by hunting for meat as chance encounters or hunting skill permitted. It put rather unique wide- ranging evolutionary stresses on the individual and tribe. That, coupled with the very early social habit of gathering food, mostly by women and bringing it back to a hearth called home seems a very likely main cause of the evolutionary path which took apeman into man. As will be noted again later, this factor, which E.O. Wilson calls the Nest was crucial to the development of Man.

 The most distinctive ethological characteristic of man is his highly social nature and skills. These, of course, are major necessities in a 'nest'. Chimpanzees and bonobos, our closest relatives, are also intensely social, especially bonobos. We will look a little later more closely into their social life for clues to our own.

  Some recent data gives intriguing evidence of how this kind of socializing scenario might have arisen. Wittig & colleagues studied wild chimpanzees and sampled urinary oxytocin levels during normal activities. (Oxytocin is popularly known as the "love hormone".) Most surprisingly they found that when unrelated chimpanzee females shared some food with another, the oxytocin levels went very high, not only in the one who gave, but the recipient also. The amount of oxytocin was actually higher than when they reciprocally groomed, which has been considered *the* most socially bonding behavior. This experiment, among hundreds of others actually, shows a bit of the biological substrate underlying particular behaviors – and even how they can be so easily manipulated. Oxytocinergic pathways are known to produce powerful pleasurable feelings in other ways too, in pair and infant bonding for example.

  Another basic question is "why are humans so excessively intelligent"? Why would apes or early man in the wild need big brains capable of performing amazing feats of creative intellect and calculus? The psychologist Nicholas Humphrey proposed long ago that the "chief function of creative intellect is to hold society together." As is pretty much obvious to us today, social life is far more demanding, unpredictable and difficult to control than any other aspect of living. Humphrey believed that it called for an extreme, unparalleled level of intelligence, and the more complex society became, the more skillful that individuals had to be to be successful. In other words, social and then cultural evolution is the main creative force in man's history, which lead to the development of our stupendous Mind. We will take this up in more detail in later chapters.

But back to tracing our evolutionary path from the first (somewhat clear) branching of the hominid tree towards hominins (Man). As noted above, it does seem now rather likely that one of the descendents of the *Ardipithecus* group was the forerunner of the hominins, possibly through the aforementioned advanced 'apemen', the genus *Australopithicus*. This latter group arose right near the end of the Miocene or beginning of the Pliocene epoch, about 4.4 MYA.

This pivotal ape-man, *Australopithicus* comprises a large and long-lasting group which radiated eventually into several species, some of which migrated out of Africa into adjacent parts of Asia. Mostly, though, they lived in south and east Africa which like now was treeless over large tracts and rather arid in some places. Some of the last, most evolutionarily advanced of those Australopithecine species didn't actually die out until about one million years ago. By this time several other hominins, including *Homo* had already arisen, possibly from one of the early Australopithecines. Opinion differs on whether *Australopithecus* was on the direct line to *H. sapiens,* but seems as good a bet as any. DNA evidence from *Australopithecus* is lacking but it is possible that this information will eventually become available and that should settle the question of relationship.

In any event, since the chimpanzee shares over 98% DNA homology with us, it is certain that we should share a considerable amount of *Australopithecus* 'blood' too.

One of the factors in the evolution from *Ardipithecus* to *Homo* was the changes in arms, including length. These lead to enabling such feats of agility and strength to throw objects with great force and accuracy which only *Homo* can achieve. This physical ability is still the greatest among any animal before or since and would obviously constitute a great evolutionary advantage in certain niches. It is quite likely that much of the meat was secured by using this endowment – not to mention fame and fortune to Bob Feller, Sandy Koufax and many another 100 mph baseball throwers. It is perhaps telling that in the whole line of the progression to modern man, beginning with the Miocene and the earlier mentioned bipedal, potential ancestors, none ever reverted back to the trees, or even as far as we know to knuckle walking. This has profound implications for why the anatomical progressions in the later progenitors became like they did. That is, there was increased selection for those traits which enhanced better survival of groups on the ground; larger brains, more gracile bones, better power grip, better hunting ability, faster long range ground speed and so on.

As one small example of physical evolution relating to mental development, there is a key anatomical feature which is strongly linked to the ability to make and use tools more effectively. It is found only in humans and has been recently found in a 1.5 MYA *Homo erectus*. This is a modification of the third metacarpal bone, specifically the styloid process, which is a little extension of the proximal end of the bone in the hand. It's function is to provide a more solid lock of this bone to the wrist bones and thereby to allow an easier and more powerful grasp of the whole hand. In any event this business of the human hand actually sheds much light on the origin of humanity.

First off, the evolution of fingers, which was a direct consequence of the bipedal habit, lead to the unique fingers we have today, marked mainly by their fineness and softness. That is, the pads are not hard and horny, but delicate and fitted with an enormous array of sensory nerves. This facilitates constantly improving dexterity through a positive feedback loop of 'better fingers, bigger brains, better survival'. The kinesthetic sense of the human finger is unparalleled in the animal world and naturally it takes a bigger brain to accommodate the increased neuron population necessary. The kinesthetic sense area for the finger in the brain is many times larger per body area than any other species. The extra brain capacity arising along with the finger evolution undoubtedly could, and was, turned to greater neural ability for other purposes.

Two other early potential human ancestors appeared in this time frame. These are the rather familiar *Homo erectus* and *Homo habilis.* One or both of these are considered to be on or very near our direct human ancestral line. Fossils of *Homo habilis* were found in Africa clear back to about 2.4 MYA. *Homo erectus* was also from Africa originally, but already by 1.8 MYA some had arrived in Asia. They were the first human type to leave Africa. Over his vast range and time span, *Homo erectus'* various radiations had led to many major evolutionarily changes, including big improvements in his tools and culture. By the time the last groups of *Homo erectus* became extinct he had a brain volume up to 1200 cc – although with a great deal of variation – and he is almost universally regarded as an early human. It is worth noting too, that *H. erectus* had a very long sojourn on the earth and lived at the same time as at least 10 other species of *Homo*, including the famous neanderthals.

Taking the whole collection of data from the ubiquitous *Homo erecti* then, it appears that he was a very accomplished tool maker, skilled hunter-gatherer and used fire. Abundant data shows many types of stone tools and also clear evidence of killing and eating large game, including even the vanquishment of a saber tooth tiger. Some of the Asian *Homo erecti* also were not above murder and/or cannibalism, as several skulls found near Peking had holes in their head from trauma.

Whether there was warfare between unfamiliar tribes that may or may not have looked exactly like them is not known, but given the evidence of territoriality, strife and war in virtually all extant primitive tribes and modern man too, it is quite likely. The later *Homo erectus* was quite clearly

22

capable, physically, of being entirely human; that is, he exhibited definite language, social structure and all the rest. Since *H. erecti* species were such a successful, long lived, widespread group, their fossil record also shows a considerable variation and changes in both their physical structure and their tools. *H. erectus* specimens have been reported as late as 143,000 years ago! As mentioned, by that time, even *Homo sapiens* and *Homo neanderthalensesis* had arisen.

Relevant to the role of warfare in the history of Man, I might note here some data from Pinker (2002) on modern mortality from warfare. He cites mortality in a variety of primitive tribes, one of which is the Yanomamo. In this Amazonian tribe nearly half of all male deaths are due to violence. Around the world the range of deaths by war range from 2% for modern Europe and America – even including the last 2 world wars – to 60% in the Jivaro tribes. The world average is about 20%. It is that high too in modern wild chimpanzees – and a number of human settlements too.

In actuality, of course, the whole human taxonomy and lineage is far from settled or clear. There are a plethora of fossil finds from all over Africa and Asia, for example, *Homo ergaster,* Java Man, Peking Man and a host of others from over 2 million years ago. However, we will not deal with any of these right now in our very truncated and broad overview of the rise of humanity.

But just to help give a feel for our own selves aborning, Figure 1 below, shows 'photographs' of some of man's earliest close relations, from *Ardipithecus* the ape(?) to *Australopithecus* the apeman, to *Homo erectus,* the man.

Figure 1. Models of a couple of possible early Man's ancestors. *Ardipithicus,* on the left, may have been the first 'apeman', while the one on the right, *Australopithicus*, is clearly an apeman (or even Man according to some investigators). *Homo erectus,* below, is considered by some to be the prime candidate as our immediate ancestor.

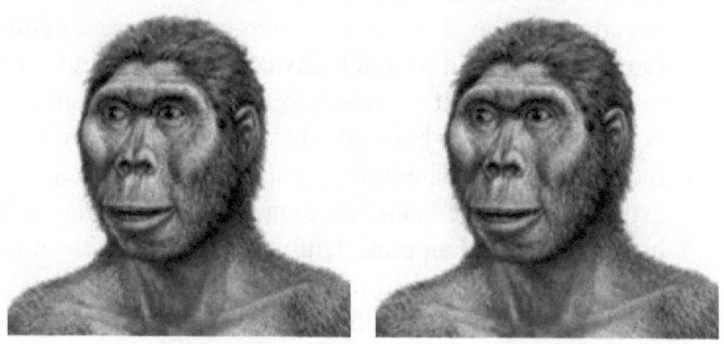

The Figure above shows that the earliest *H. erecti*, while clearly humanlike, were also clearly a long way removed from us. Their skulls were low and quite flattened, with pronounced brow ridges and prognathism – protruding snout. Their brain size early on was on the order of 800? cc and the tools found near them were of the most primitive type. The latest fossils, less than 500,000 years ago, however, show them with much more modern form, with very advanced tools and a brain size of about 1200cc, just a little less than those of the first true *H. sapiens.*

Here I might interject the saga of a very extraordinary event which early on staggered the rise of man in Africa especially. It serves as a cautionary tale for us and for all time. About 74,000 years ago a super volcanic eruption, the largest in all hominid and hominin history, occurred in Indonesia. Ash and clouds covered the earth for almost a decade, and the climate was changed for over a century. Obviously, it lead to much devastation of the earth's mega fauna and flora. It is known to have sent the human population everywhere into a severe decline, lasting nearly a thousand years. The human population in Africa is estimated to have been less than 20,000, and may actually have been less than that. Humans came very close to extinction. That should tell us something about the sometimes very tough times that our ancestors had to pass through. Struggle for survival could have imposed some pretty stringent selection pressures to stretch their – our – wits.

Two last major actors need now to be introduced into the human lineage. These are *Homo heidelbergensis* and *Homo neanderthalensis.* Many anthropologists suspect that *H. heidelbergensis* was on the main line to modern man, although it is possible that he simply coexisted with other species, such as *H. erectu*s and possibly even *H. sapiens.* A recent suggested scenario is that *Homo heidelbergensis* arose in Africa, probably from one of the *Homo erectus* groups and is in reality *Homo sapiens'* immediate, direct ancestor. It may be worth emphasizing that while *sapien*s was born in Africa he subsequently left Africa at least twice, once about 2 MYA and again ~50,000 years ago.

The last of the 'old time humans' we will consider is the Neanderthals. They have gotten kind of a bad rap in the press and academic world ever since their first fossils and artifacts were discovered. The Neanderthals arose in Europe about 300,000- 500,000 years ago and are thought to have

25

arisen there, rather than Africa or Asia as all others did. Although their faces and bodies were cruder in some respects, their brains were the same size as ours and they produced quite modern tools. They clearly had language, culture and thought processes much like us, although in some ways they never seem to have progressed much either.

Recently the Neanderthal genome has been completely sequenced, although analysis is in its infancy (see Paabo, 2014). About 1-4% of the DNA in modern European humans is Neanderthal, indicating interbreeding. They were light skinned, some apparently even with freckles! It was reported too, that possibly about 1% of Neanderthals had red hair. They were quite short and stocky in build (they lived in a cold climate) but had the largest brains of any hominin, including ourselves. Their – still stone – tools were the most advanced of any until our own modern group surpassed them. They remained however, always hunter -gatherer tribes. In their case their diet was primarily carnivorous, unlike virtually all other early hominins. To reward him, his successors, us, probably helped wipe him out!

The extinction of the Neanderthals is really not a mystery. In competition with the new kid on the block, Cro-magnon (*H. sapiens*) he had no chance. In the long run though, *Homo neanderthalensis* did not die in vain. While they were out competed and also sometimes murdered, they also often enough mated with us. As noted, we have incorporated about 4% of his genome into ours at this moment.

This last kid on the block, then, is us, *Homo sapiens sapiens*. The earliest fossils of us – Cro-magnon man – are found in Europe beginning at least 100,000- 50,000 years ago and had every last anatomical characteristic of present humans. Neanderthal's as well as Heidelberg's skeletal features were really pretty close to Cro-magnon so while the missing intermediate types might ultimately be found, there are no mysteries to be solved there. While earlier homo species had almost all the essential human features and

26

mental powers as we, including art, superstitions and death awareness, Cro-magnon man had them in spades. Their tools also advanced in number, skill and inventiveness by leaps and bounds. From an estimated 12 tool types of *H. erectus,* Cro-magnon had at least 60. Not only were they more utilitarian, but some, especially their bone tools included artistic embellishment. More interesting still is that their tools were continuously changing, more in a century than in millennia before.

Thus, when exactly *H. sapiens* first arose is still not at all clear. But it does seem to have been in Africa about 100 to 300 thousand years ago. *H. sapiens* mitochondrial DNA from the first so-called "Eve" was found in a 160,000 year old fossil. The Y-chromosome (the so-called "Adam" DNA) of *H. sapiens sapiens* was first found in a 300,000 year old fossil. The evidence seems to be mounting that *H. sapiens* arose earlier than long suspected, probably somewhere around 300 thousand years ago.

In any event, so here, somewhere in this progression we have him! Us! The physical changes from even the ape-human transition forms like *Ardipithecus* on through *Homo habilis* or *Homo erectus* and the rest, are really not all that great or noteworthy – – except for the tremendous increase in the size and complexity of the brain (especially the cortex). Over the whole play of prehistory the main physical changes – which I have very briefly sketched – are well documented in numerous books, films, texts and articles. While brain size is the main outward change, the whole skull was revamped, mainly the upper cranial vault. The face flattened, became smoother, less thick -boned and the foramen magnum shifted completely under the cranium. Along with the arm shortening, bones in general became more gracile. The hand and also the foot underwent subtle but biologically significant changes.

By looking at a chimp and then in the mirror, a good estimate of the nature of the changes can be had. Bipedalism and adaptation to a new way of living on the ground in highly social groups appear to have been the main factor in both the physical and mental evolution.

Of course, the main issue we face in determining where, exactly, in the progression which goes in general from the Ardipithecus type prehuman to modern man do we pronounce him fully HUMAN. Is it, as he goes through (or maybe passes nearby) an *Australopithecus*? Or *Homo habilis*? Or one of the *Homo erecti*? *Heidelbergensis*, or maybe a Neanderthal? Criteria for inclusion into the human fold varies, from a simple "bipedal, tool using, conscious ape", to fully culturally advanced, even artistically sophisticated, modern Man. For our purposes it doesn't really matter exactly what forebear and when exactly we are generally willing to declare him Us, as it will probably never be beyond appeal. It will likely be forever impossible to stamp this one, or that one, with the honor of being the First Human, because there will forever be argument as to what is a human anyway.

Few authorities are anxious to categorically state even their best guess. Lewin (1982) rather timidly, but at least publicly points to a time 300,000 years ago when it appeared that a *Homo erectus* may have passed the bar, as there was a family shelter, carefully arranged and they seemed to have used ochre to draw symbolic designs on themselves and on rocks. If this turns out to be actually true, then they might very well be the originator of a new, higher level of consciousness and Mind.

What is a reasonable taxonomy of the brain/mind, and what, exactly, are the requirements to be a man? It might be useful to make a kind of list and then briefly try to sort things out. Consciousness? (I'm using the term here as commonly understood by scientists – or even common parlance.) Yes, of course. Consciousness is a major characteristic, but all sorts of animals and all mammals have a goodly supply too. Recently, Peter Godfrey-Smith (2016) made a very

28

interesting, if not compelling, case for high levels of consciousness appearing in one of the very lowest phyla of animals which arose over 300 million years ago. The phylum Mollusca contains the interesting Class "Gastropoda", which includes the snail-like *Nautilis,* but more particularly, octopuses. These lowly animals show amazing levels of consciousness. 'Curiosity', and 'friendliness" are some of their surprising 'personality traits.'

So, back to our list. Self awareness? Surely, but other animals, especially some apes too, have a fairly high level of self awareness. Obviously though, ours is a higher and unique mental achievement. We will come back again to this very critical factor.

Intelligence? Of course, but again a great many animals possess a quite high level of intelligence. Birds and a great number of animals exhibit sometimes an astonishing level of ability to solve problems; even to the higher instances of thinking out a solution. One example recently shown on National Geographic was a film of a chimpanzee presented with a peanut at the bottom of a tall thin glass cylinder. He tried first digging it out with fingers, but the glass was too tall to reach the peanut. After just a few seconds of looking at the thing, hitting it couple times, and obviously puzzling over how to get at the desired treat, he ran over to his water bottle, sucked up a large mouthful of water and spit it out into the glass. After the first attempt he saw that the peanut came up a little ways but he still couldn't reach it. So he repeated this 2 or 3 more times until the peanut had floated up so he could get it. There are hundreds of other documented cases of similar, or higher, intelligence in numerous species, including elephants. Obviously, across the animal world there are large gradations in the extent of intelligence (even in humans, of course, there would appear to be a – rather limited but still very, very high – range of this mental capability).

Language and communication? These are obviously central to consideration of the higher powers of man. It is also very true however, that communication is not at all limited to man, some animals perhaps even with primitive types of 'language'. Nevertheless, true language is unique to man, and ideed is the prime mover for humankind's origin. As has been shown recently by evolutionary psychologists, language could be a *result* not a cause of man's unique mental state. The 'competence' to use language to borrow Dennett's term ("From Bacteria to Bach and Back", 2017) may have arisen, by Darwinian evolution, in early prehumans, but language itself belongs to man alone and is the critical factor in making him what he is today. Once language arose, in fact, cultural evolution has trumped everything and took its own rocket course, which is accelerating still.

But the sine qua non of being human is really our Mind, an inner Me as the Controller and the Knower. This ineffable composite characteristic, which allows us to call, and know ourselves as a distinct entity -- Self. This consciousness includes the ineffable factor of reflectiveness (Comprehension is the term Dennett, 2017 uses). We think; and we know; and we know that we know, and I know that you know. This makes all the difference in the world. This feature provides the machinery and fodder for all of the inner workings of the mind. This produces all of the thoughts, ideas, art, religion, science, predictions, joys, artifacts, governments, etc. etc. that we know to exist. This makes Man special, unique. This is Man. This also makes Man the master of his universe.

The Mind is our internal monitor and motor, which semi-magically splices all our external and internal inputs together into our own continuous story. *It* talks to Me, and I talk to *it*, and together our life story, instant by instant is fabricated and made into one. Adam Gopnik rightly says our Self, our mind, is made up of myriads of these stories---- and furthermore, we get to make them up and string them together as we instantaneously determine. To me, the essence of Mind is that part of Self which makes up our story of living; each moment is our Now—our very existence-- and moment by moment we stitch together a more or less continuous story of who we are, what we were and are, what we think and feel etc. This Knowing Nowness *is* Me over time. The neuroscientist Siegel (2017) among many others, emphasizes this feature, which he says is the major part of the Mind, or Subjective Reality.

Many other current neuroscientists, like Antonio Damasio, 2010, and Michael Gassaniga, 2018 and even philosophers (Daniel Dennett, 2017) give thorough and compelling expositions of the evolution of mind. The title of Damasio's book gives the gist of it; "Self Comes to Mind". The origin of this knowledge that there is a true Self in the Mind (which makes us Human) came about piece by piece in the evolution of the higher animals, beginning already with the reptiles. It arose and developed in complexity from a 'protoself' in animals like dogs, elephants and a few others, but it fell well short of true knowledge of the 'owner' Self until our man, with his language, arrived. This is the incredible and magnificent line we alone have crossed.

The fossil record is pretty shaky to pinpoint just when these characteristics arose. It is common nowadays to place the first truly fully mentally loaded modern man at least 70,000 years ago, when considerable evidence of his potential powers exists. It is quite likely we could move the critical date back maybe a hundred thousand years or more, but who will ever know. Whatever the exact date and hour, he then was *Homo sapiens*. If we were Men then, we must have had morals then too. The fact that moral intuitions are rooted in biology is an immensely important concept. We need not learn that cruelty is wrong from religion or a tutor, if we don't have an inkling inside.

Now that we have come to the clearly man-only aspects of life I should note that the current state of man, as of the whole planet, is not fixed. Evolution is proceeding just as it

31

always has – very slowly and imperceptibly to be sure. In fact, the current state of evolution of man's magical powers is still in an early state. Both mind and morals can continue to evolve seemingly at Will, and one may hope, rapidly at that.

How Might We Have Come By Our Mighty Mind?

In the beginning, a newbie to ground living, living in small bands and having to forage for food and avoid predation, walking upright served to his advantage. Early on he had a territory that he knew and wandered freely on. He *knew* where the potential food was or *predicted* to be, and this was maybe his first big advantage. At any event, from his own traditional home space he saw the lay of the land even beyond his territory, so exploration and wandering was probably attractive from early on. It is thought that a typical African savannah territory was 5-10 miles in radius, although almost certainly they explored larger areas. They would sometimes came upon another band or tribe, too. At this interesting point sparks might have flown; assimilation and/or war often probably resolved things.

All of which kept the pressure on to develop more and more his selective advantage of 'using his wits'. Big brainy guys bred bigger brained descendents. "Collective learning", for example was probably a significant factor. Collective learning means that somewhere along the line, probably quite early on, the whole tribe became involved in group learning. That is, each member of the tribe was part of the 'chain of knowledge' (i.e. knowing the territory, food sources, social norms, etc.) and this to some degree passed to children and on down the line. Evolutionary success is more dependent on 'information' and how some individuals in their group used their brains so that the whole group was successful, i.e. social evolution ruled almost from the beginning. Cultural evolution, caused by our unique evolutionarily advanced brain is the forge by which the magic Man arose and thrives. As mentioned, Damasio, Dennett, Gassaniga, Wilson (and virtually all current authors) agree and provide strong, detailed and beautiful expositions of the salient evolutionary features involved. It is more than a little complicated and will require a couple more centuries to sort even the basics of this most important (last?) scientific frontier.

By the way, I might highlight a couple of obvious facts that stare us in the face and yet remains below the mainstream radar; Man is one of the youngest animal species to have evolved, and also the most widespread – ruler of the world in fact. He got this way by specializing in the Mind. Mind is evidently the most powerful force, as well as most mysterious ever derived.

Anyway, the band or tribe – probably typically 25-100 people – was pretty much on their own and the individual lived or died largely by the success of the tribe. Tribes were actually rather few and far between, as the population of early humans probably was only around 50,000-100,000 total early on and probably less than that in Africa for a million years. So day after day, the group carried on the daily grind, gathering and hunting food, taking it home and eating it. Having a home base as the social focal point of all activities has been proposed many times as the crucible in which some of the physical, but almost all of the social evolution took place. In fact this very action lies at the heart of the success of the human species. This is where language, becoming ever more and more complex probably originated, to gossip, to make plans, or argue and the like. This created a positive feedback system which basically continues today.

In any case, the social group was from the beginning the essential unit of our storied human evolution. That meant that while the innate traits of each individual were a little different each from the others, it was survival of those traits which enhanced the wellbeing of the tribe that was more

33

important than the survival of any one individual. 'Getting along' with the rest of the group was undoubtedly a major factor, as it is today in virtually all mammals. Being able to 'read' other pre-humans' intentions would be a major job too, as it is today. This takes a big, fancy brain, which also, critically, drove, and drives cultural evolution. This is the essential nature of the social evolution that went on, unnoticed and unknowable down through the generations. This is also how and when morality evolved, some more aspects of which we will take up a little later. Obviously it was successful and lead to the final genetic innate traits that we had over 100,000 years ago and that we largely have, for good and ill, still today.

Harvard Professor E.O. Wilson, the world's long time authority on evolution, has especially advocated that this early human practice of having a NEST (as with ants and bees) was the crux of selecting for larger and better brains and thus leading to the lone path to modern man. Wilson (2012) argues that (excluding the oddity of naked mole rats) there are only two major groups of animals, the social insects and man, that exhibit true sociality – eusociality. In both, the nest is a pivotal feature of the society. (However, Wilson would probably agree with Pooh "You never can tell with bees".)

Basically, this concept is simply that the group has a fixed home base, in which the reproductive unit lives, and in which some of the group goes out regularly to forage – and also most significantly to hunt. Also significantly, they bring the food back to the hearth to share and eat it. Only a few species in the Mammalian Class, especially primates, lioness's and some dog packs, African wild dogs and wolves to some extent do this. This by its nature requires some division of labor and thus more and more emphasis on cooperative effort for the survival of the unit. This is the roiling and agonizing crucible in which our own current behavior and morals were forged. Getting along with other

34

tribal members was a Must. That takes brains. In a later chapter we will again come back to this important aspect, the evolution of the human nature.

Modern Man Into the Holocene Epoch

This brings us down to the end of the Pleistocene epoch and the beginning of the Holocene, the modern age – – when *Homo sapiens* started to become sapiens squared! Even before our now modern selves, *H. sapiens sapiens* had arrived, a goodly number of social advances had been invented; big ones, like language, myths and religious-like rituals, art and so on. So even well before the end of the Pleistocene, the capability for enormous progressions in social living particularly, was lying latent. From that time, each improvement in some – cultural – aspect lead to geometrically feeding off it into more and more, until it quickly reached an avalanche. Cultural changes are still avalanching, even without any substantial physical changes in brain or other features.

That 'final' explosion was a mere 12,000 years ago. His rapid physical evolution was over but his social life and social evolution was about to explode. Time to turn the page and ratchet up the ante on Mentality and its products. The young *Homo sapiens sapiens* had already inherited the results of all the heartache, pain, triumphs, joy, bumps and bruises, and ego deflating or inflating dramatic and boring day to day life and survival of all his predecessors for a couple million years. Inside our quite beautiful looking and adapted bodies, particularly our big heads, we carried mainly the triumphs, but many of the defects too, of all the preceding generations – perhaps about 350,000 of them. Later, we will look at some of the latest scientific information about possible neurobiologic, psychological and social and evolutionary mechanisms that might underlay some of these inherited characteristics.

At any rate, by 12,000 years ago, humans had already established life's basic routine of social living and their place in nature seemed secure. In particular, by this time in their history no matter really where one might like to pinpoint *the* point at which we were truly human, this Man had to have lived by his wits. In fact, this is what made him human in the first place. They had already, although no one knew it or would have thought of knowing it, circled the globe through their successful lifestyle; the first species ever to have spread worldwide, at least on their own merit (not like rats, for example which piggyback on man for their universal distribution). The simple fact that the earth is spherical has caused us to have met ourselves on the ends of the earth and everywhere in between. This now constitutes mankind's most difficult problem today, with unsustainable populations, dangerous environmental changes and oftentimes explosive social tensions. Solving these will require adaptability by our magnificent stone age brain and also, importantly, assuming responsibility for the planet---- and adopting plans and programs by use of hard won knowledge, experience and science

CHAPTER II.   CIVILIZATION STARTS HERE

Having traced our lineage down to about 12,000 years ago, we can now see that obviously there was another 'Turning of the Corner' at this point of time. Agriculture began and the first villages appeared. These events mark the beginning of the modern world, with settlements, fire to harden pottery, copper smelting, agriculture, civilization, rapid technological innovation and all the rest.

But more than 5000 years would still elapse before the first writing began, indicating how really close to the present is the period of time that we normally think of as 'real' history. Only about 300 generations of our ancestors separate us from that hoary period when writing was invented. Virtually everything that schoolchildren learn about the world has occurred since then; everything that man seems to have invented, written or thought occurred since then. Not really of course, as Paleolithic ancestors obviously also had thoughts, and even expressed many of them, but what, exactly they were about is not knowable.

So back to 12,000 years ago, when all this evolution had produced the man we are today. All the mental tools were present and what we see beginning now is the initial working out in fast time their sort of pent up inventions, morals and socialism. Let's say the year is 11,999 BC (12000 years ago today). Tribes of the 'traditional' *H. sapiens* are now spread out all over the world – even though the trickle into the Americas started just a few millennia earlier. People numbered all- together possibly 500,000 to a million individuals worldwide. Some groups have domesticated dogs and most have the very latest models of beautiful stone tools and nice wooden spears. There was beginning to be more variation in tools, pottery and local habits between mountain ranges and valleys than had been between continents just a quarter million years earlier.

Now let us speculate on how it may have happened and some of what actually is known to have happened around that period to have created a memorable – at least in retrospect – moment of extreme change. At some point that 'year' – or within a few centuries – in the middle east, or maybe in southeast Asia, perhaps a couple of adjacent tribes living in an auspicious fertile valley 'decided' to stake out a expanded common living area and really settle in. They moved into more permanent shelters of wood and earth and stone. Already, people were inventing the idea of becoming a stay-at- home species, which of course, meant eschewing the old ways of hunter-gatherers. Why?

We may never know, but in any case, in a few especially felicitous locations along rivers, fertile ones with abundant native grasses, villages had already sprung up where a somewhat expanded tribe founded a settled existence. It would have been a rather easy life really, with abundant game and harvestable wild barley, wheat, nuts, and berries readily available for the foreseeable future. 'Gathering' then began to morph into 'harvesting' which increased and eased food production. Maybe one of the women in the now expanded group had noticed that some wild grass seed that had dropped near the home, had germinated and grown into nice barley or wheat that they had simply been gathering for decades. When it was hand harvested successfully, it might have made sense to them to next year drop some more of those seeds around to see if it would work again.

With more and more of these settled-down villages, civilizations could not be far away. With more settlements came larger and larger villages and closer contacts, which in turn must have required greater social structures and mental explorations. The chiefdom concept arose early from these settlements; the idea of a large group acting under a single man – and possibly purpose. Before, in all the twists and turns of prior tribal existence, natural leaders undoubtedly arose from time to time, but there hadn't been a central 'brain' of the entire organization over large territory. Sinple tribe organization was probably the highest level of 'government'. Ever since, though, for better and worse, the 'brain' of larger groups has essentially been government of some sort. It is important to state again, that the *ability* for mental exploration and to conceive additional and more complex social structures had been already bred into the species unwittingly in the previous thousands of antecedent generations which had gone through the fire of natural selection for eons.

In any case, wheat and barley and also peas are known to have been wild harvested and domesticated in several places in eastern Turkey, central Asia and Iran by this time. Domestication and husbandry of various animals too, like (besides dog), cows, pigs, sheep and goats is also known to have occurred in this period. The first known domestication of cows, for example was clearly accomplished in the Fertile Crescent by about 10,500 years ago. Sheep, derived from the wild mouflon, were also domesticated about this same time, first in Mesopotamia. The wild goat, widespread in Asia and Europe had been hunted for millennia by Neanderthals and other men, and had been tamed also in the Mideast by 10,500 years ago. Along with the animal domestication there must have been too, some kind of 'farmers'. They were the ones to invent the next big thing in the world – – agriculture.

What a momentous act that mythical woman made somewhere in that millenium! It would be interesting if we knew what her name was, to pay homage for that simple, fateful act(s). That was the moment of the First Wave. That was the moment(s), when Agriculture, our chief enabler and underpinning was invented and made modern man feasible. And also the Second Wave – the Industrial Revolution – , and the Third Wave too, our own Information age. With agriculture came permanent settlements, with many more and larger and better houses, technologies etc. avalanching directly down to our own time.

One of the earliest large villages made possible by

39

agriculture was 11,500 years ago in Syria. This was a fairly small city, but amazingly also had a community center. This was a large mud building 22 by 19 meters and had a large central space with many big benches around, indicating perhaps the first 'political' organization. Around this were a number of small houses. Already the people there raised cattle and also harvested wild wheat and other indigenous crops like flax, emmer, peas, barley, beans, lentils and fig trees.

Fibers and cloth made of flax and/or cotton, etc. were woven in this era too. In China, silk textiles are known at least from around 7500 years ago and felt fiber is even older than that. Not long after, even larger villages were found, especially in Turkey, some of which contained hundreds of well- constructed houses and thousands of people. Similar developments occurred all over the near East and China too at roughly this period. These same kind of settlements and villages arose also, although a bit later, in Egypt along the Nile around 8000- 9000 years ago. The agricultural phenomenon reached central Europe about 7000 years ago by immigration from the east.

Along with all these developments, various other high tech activities, like advanced pottery kilns, were also attached to these early villages. Many villages carried on substantial trade with rather far flung areas for such things as beads, timber, some minerals and other small items. Trade had actually started in the first tribes of *sapiens* in Africa as long as 130,000 years ago.

All in all, these advanced villages, cities really in some cases, required considerable advances in organization and inventions. These early villages and cities must have solved the inevitable knotty social and organizational problems, as there are no archaeological indications of chaos, mass killings, or disintegration and they lasted for hundreds of years. The concatenation of new horizons and synergy between expanding social structures and conscious, expanding thought and more 'stuff' has been the story of man ever since. This kind of exploding progress – i.e., cultural evolution – is never ending and ever geometrically expanding the whole time, very little less then than now. The more material – 'dough', or information – one has to work with, the more opportunities for incrementing it there is; a true positive feedback situation that as far as we can see is to continue forever – for better or worse!

 We needn't go too deeply now into details of these last 12,000 years because it is familiar, in outline at least. Thousands of books and lectures and scholarly explorations continue to pile up practically daily. For all practical purposes, everything we now know, use and understand were introduced since then. It is important to state again, that while all of humanity's high accomplishments occurred very recently, we must continue to understand that the physical underpinnings of the human condition had already been laid down by the long (more than a million years) prior evolutionary history. All of the potential brain power and passionate beast had already existed, although not yet fully unlocked. (Nor is it yet; we are the 'young-uns' of the planet.)

    All groups carried with them all of the physical and some of the cultural baggage, such as it was, from our early and late progenitors. Even in the New World, starting essentially from scratch with stone age tools, man had reached South America, reinvented agriculture, cities and civilizations. Everywhere, all around the world was Man; big brained, 'cultured', and with an active mind which already 'knew' how to write orchestral symphonies, do space travel, nuclear physics, the Internet, make Dumb and Dumber movies, rain down genocide from the skies, mud wrestle, mutilate new brides or children, gamble, declare war, invent new religions or myths, or just plain think up stuff of any kind at all, even crazy stuff, and all the rest. In short, he already had both the engorged prefrontal cortex as well as the amygdala and lower brain structures already loaded with all of the reflexes, deep set emotions and instincts which had been hard won the

41

previous three million years. These neural structures and at
least the predilection to respond to various environmental
stimuli in programmed ways, he took with him – and still
carries – into the next phase of human development.

This brings me to emphasize two very important points; (1)
What man was 12,000 years ago, he basically is today still,
without much evolution in physical structures or in those
inculcated neural activities which had locked in certain
behavioral predispositions; and (2) everything else about
him has changed through cultural evolution, i.e. his
environment and his psychological and mental inputs and
through-puts (in degree only, however).

OK, now we have traced humanity down to about 6000
years ago, or 4000 BC. The years following 4000 BC marks
another watershed in the history of the world. Man invented
civilization. He already probably had civilization, but didn't
know it then – or sometimes since. Civilization, it is said, is
basically mostly the trappings of high strung Man, meaning
various things to various people. It is sometimes defined as
the development of an urban conglomerate, formed by a
highly organized class- based society and having
monumental architecture and written language. Some define
it loosely as large groups of people who have highly
organized culture with an intentional method of governance
(cf Brandon, 1970). Writing is a prominent feature of most
civilizations but not all; the noteworthy exceptions are the
Americas, where the Incas, Aztecs and Mayans never
developed a true written language.

Civilizations may probably best still be defined as the
organizing of city-states or agglomerations of city-states to
form some kind of national government or empire.
Civilizations are entirely the mental inventions of a large
social group and are the end result of a great deal of fast and
furious physical and social evolution over less than two
million years; plus frenetic cultural evolution over just
12,000 years or so. For example, the first civilizations
invented writing, the wheel, government, the Bronze age and

a thousand little adjustments to social living almost right at the git- go.

Just a word here regarding evolution of man's culture. By that I mean, as do most scientists, those changes in thought or other higher mental activity that occur in an individual but which are transmitted by culture (i.e. 'information') to other individuals or groups. The basic idea is that changes in behavior do occur over time. These are (mostly) NOT genetic changes, although some genetic changes undoubtedly occurred sometimes. One example is a mutated gene for lactose tolerance, which allows even adults to drink goat milk. However, over the next centuries many changes in individuals and groups will evolve, mostly as a result of gene-culture co-evolution, as we will explore a bit later.

At the outset of our tour of civilizations I might also note the obvious; there have been very few civilizations in the history of the world. The number depends on the particular lumper or splitter who catalogs them. I lump them into ten, one of which is our own. It is rather hard to argue that there is more than one at the present time. In the beginning though, there were four, each of which arose independently. Talk about great minds thinking alike! It is also very telling that three other civilizations arose, each independently, in the Americas a thousand years later.

It is probably not accidental that these great, really unprecedented and unique human events, did arise at different places at different times by different people. It speaks to the ingrained, evolutionarily determined hard wiring of the brain to follow impulses which seek to develop large collaborative, cooperative, social networks to accomplish what are perceived as reasonable enterprises. We casually call these inborn tendencies to respond to certain environmental and social conditions in certain, general ways "human nature". We do not imply, nor at all mean, that these responses are preordained, or inevitable, or unalterable, and not without an infinity of possible modifiers or truncation.

43

Just that stuffed into all of our common neural parts, like cerebrum, cortex, hypothalamus, cingulate cortex and their complex modules and activity, there are genetically determined ways that their neural circuits will respond in similar broad fashion to specific stimuli.

What I would like the reader to keep in mind now as we survey recent human history – a short history of civilizations – is a mental image of our earlier hero, Man. And what our biological heritage has wrought, arising first from the muck or ancient seas, progressing through to our pretty mammalian friends and finally the fledgling Man, naked and alone on the plains of Africa. How amazing is it, that this puny upright beast could come so far, so fast, and then enter this strange new world of half nature- half gigantic built structures! How did Man somehow will himself into this massive global all-powerful behemoth, lord of the planets?

As Stephen Jay Gould cogently argues about how natural evolution could give rise to such high mental powers " Our large brains obviously originated for some set of necessary skills in gathering food, socializing etc, but these skills do not exhaust the limits of what such a complex machine can do. A computer built to issue paychecks can also manage all the company's accounts, solve differential equations and play a passable game of chess." Also, Carrington, in his interesting book, "A Million Years of Man" says " Man, like other organisms, is engaged in a constant process of adjustment to his environment and the breakthrough to the new level of biological organization represented by civilization is another example of evolution... Moreover, these principles are constantly changing with the physical and psychological environment similar to that of a biological organism, or it will likewise become extinct by a failure to adapt."

History Begins: The Way Things Were, 4000 BC to Rome

But back to around 4000 BC and the inventions of civilizations. In Iraq (Mesopotamia) and particularly the more eastern and southern areas (Sumeria) there were dozens of large cities of 10,000 or more along with more than a dozen smaller, but still autonomous, additional cities. Each of these apparently had a city-king so the concept of a king in the scheme of governing such a large new structure had already arisen. None of the names of any of these earliest kings has come down to us, as writing had not yet been invented – in fact, we do not know the name of a single human person over the whole preceding time. But some time before 3500 BC someone, or some group in one of these cities produced a type of crude written language. This earliest writing was really proto-writing but it fairly rapidly developed and by 3000 BC 'real' written language, cuneiform, was being used in Sumer. History, we could say, begins here. That is only 150 generations removed from us.

These Mesopotamian cities large and small, had common public structures, along with houses all around a ziggurat or shrine. The houses, most of mud or mud bricks, were jammed together to form basically a housing development. Early on, many of these cites did not build walls around them, but that practice changed within a few hundred years. Agriculture, as ever, provided the base of the economy. In any event, this loose group of city-states created a civilization. The first one ever, all by itself.

Shortly after the year 3000 BC some 30 or so of these still-growing city-states in this region consolidated under one king, the first of which seems to been called Enmerkar, and the second, a famous one, the legendary Gilgamesh. This, the first of the Mesopotamian nations, was called various things at different times but Sumer sticks. That nation, under indigenous rulers, didn't last too long. In about 2350 BC a king, who is best known as Saragon took over and established his base at Kish. There doesn't seem to have been much of a war, as not much is known about how power was wrested and there was very little change in the lives of the people. In 1972 BC, however, another power grab from another near neighbor, this one based in Babylon which is right near the heart of Sumer, occurred under an able governor, King Hammurabi. He and successors' held sway for over a thousand years.

The Egyptians meanwhile had been creating one of the first civilizations and the first true Nation. It is again noteworthy that this feat was made from within, another example of parallel social evolution, which – as is true of some of biological evolution too – occurred over and over again. The Egyptian civilization, at least on a national scale, can be said to have begun in 3150 BC. Then, the first King (who also has the distinction of being the first human being that we know the name of) took control over and unified all of the scattered city-states along the Nile. These flourishing agricultural city-states had been long established, with a high culture. The larger ones had a city chief. Somewhere along the line around 3200 BC someone had invented hieroglyphics. This is how we know the name of "Narmer", the first Pharaoh of the longest lived civilization in history. But after that we know the actual names of lots of people and almost all of the succeeding pharaohs, all the way down to the last one, Queen Cleopatra, 22 BC.

We needn't trace the progress of the great Egyptian civilization in detail, as it is well and generally known. It should be noted that their hegemony lasted 3000 years. While much of their reign, especially the latter third was marke6d by the more or less usual "civilized" activities of war, strife, and disaster, overall the rule of the Pharaohs was fairly good. They early established a good deal of local autonomy and the official structure involved advisers, councils at several levels, governors and an Army, plus an army of well-regulated civil servants. They had public works

46

departments, mail service, a tax department and several other such public units. The government was rather benign, pacific and based on a moral code. Much of their social life was centered on their religion.

Their social structures were allowed to flourish as they might in the various regions along the two thousand miles of the Nile. For the most part, people were treated rather well, except for slaves and other lower classes. (As the song says, however, "When Pharaoh's around, you get down on the ground".) Almost from the earliest, they began to build and invent things; not long in fact before the great monuments, leading to the pyramids. Their art and architecture was outstanding from first to last and constitutes a strong influence on all subsequent civilizations. They made the first numbers and many advances in science, mathematics, astronomy, chemistry and medicine. For example, they pioneered the use of the first effective pain killer, opium. They explored mines, used lead, and smelted copper early on. They invented ships, paper and the calendar among other things. And, of course, the magnificent library at Alexandria.

At the end of their civilization, essentially at the junction of the Roman and the beginnings of our own (Western) civilization, the Egyptians had not evolved a jot physically, nor in mental capacity, nor in any of their neural hard wiring and emotions. The only visible thing that changed was due to social evolution. Social and psychological capacity changed more and more rapidly as the years went by, just as in the other civilizations – this *is* social evolution. What they developed and invented are pretty much all now ingrained in our own individual selves through learning and also into every society. They are a part of us, though not a part that shows except mentally and socially. The ideas that they first thought of, or at least put into either words or material objects, plus the extra ideas and thoughts that these generated in us and our predecessors are also now part of us.

Two other civilizations developed independently during roughly the same time frame. The first of these was the Indus civilization, which developed about 2600 BC, first in the city of Harrapa along the Indus river in now Pakistan. A great many quite large cities were built, extending over a large area almost to India. It is estimated that the nation had a couple of million inhabitants. Even before these cities began they had had a thriving agricultural society, but their main feature early on and throughout their civilization was their remarkable urban engineering. Their cities were very well built of wood and mud brick, with walls around. Everything seemed to be built according to a master plan for the whole city. They had many large public buildings, but no temples, shrines or monuments, which is unique in any civilization. The cities were built with advanced water and also, uniquely, sanitation systems.

Of note also, their far- flung cities were autonomous, as there were no overall kings, and they had no armies, or priests either it seems. Each city had a ruler and a well- run civil service and in general had efficient and rather egalitarian government. Religion had a very light footprint as there is no evidence of public religious rites. They did have a small number of gods which apparently were carried over from much earlier agricultural traditions. The main one was the Mother Goddess and then some minor deities to various plants and animals. They invented their own unique language, which is called simply Indus script. This, however, died out and is not totally known to this day. They also invented their own very sophisticated system of measurements, as they measured length, time and weight very accurately, which is no mean feat to sneeze at.

Around 1750 BC the Indus valley and its inhabitants witnessed the beginnings of a couple of momentous events – one in the form of a natural cataclysm and the other a social tsunami. Both of these profoundly shaped the society of the subcontinent into what it looks like to this day. First off, the

48

climate turned cooler and drier, and since each city-state depended upon river agriculture, their cities began to suffer. There began a slow decline in the indigenous population. At the same time there was a vast immigration of people, called the Aryans, from central Asia. By 1600 BC the Aryans had largely supplanted the Indus culture, ushering in the last of the true Indus civilizations – the Aryan-Indus civilization. This 'takeover' was accomplished mostly bloodlessly and the combined Aryan-Indus culture went on to become largely the present Indian culture. The Aryans brought with them the Vedic religion which essentially morphed into Hinduism, much of which continues today.

The last of the ancient national civilizations was the one that the Chinese developed. It began and remained unique and independent for a very long time; indeed it has remained more or less continuous down to today. The Chinese had been isolated for a long period and 'had to' independently invent Civilization over again. The path they took to civilization followed basically the same path that most of humanity had followed; that is, they independently re-invented agriculture, villages, better tools and all the rest, very similar to what all the other civilizations had come up with. The timing of some of these parallel evolution events was a bit later than elsewhere but smelting, the wheel, written language, etc. developed just as it had earlier. Curious.

Briefly summarizing the Chinese experience, we see that the first real record of a true Chinese civilization is found in settlements along the Yellow river around 1600 BC. The Dynastic mode of governing was the first type and remained virtually the only type until modern times. Early Chinese religion also developed differently, and while little is known for sure, it was certainly unique. Rather elaborate burial ceremonies are the best documented, as are some of the folk religious activities, which were built around homely objects and events, like harvests.

49

Perhaps the earliest organized Chinese religion was practiced by a special class of "Diviners" and was called Divination. They used cracks in various dried bones to predict (probably more like proscribe!) such things as ceremonials, tribute, orders, new settlements and many other city events. Only much later did a dominant religion arise, in ~500 BC with the birth and life of Gautama who introduced Buddhism. Their development of art was also of their own distinctive kind but it has been extraordinarily rich and skillful from early on. The Chinese civilization was obviously different, but original and the basic inventions and progression in thinking was similar.

I might note here that of the ten major civilizations which have existed we see that seven arose from and by indigenous peoples pretty much from scratch. These include the four surveyed above plus the three in the Americas, which also developed independently and in their own peculiar ways. I will not survey here the Inca, Aztec and Mayan civilizations but will look just at the last and most recent ones, i.e. The Big Three: the Greek, the Roman and the Western civilizations. These obviously were not invented de novo but relied on much borrowed information and technology. The key words here are "information" and "technology", as these most recent civilizations relied on both, not only to build up the civilization, but even more importantly to develop the institutions and methodology to continuously modify and improve the whole enterprise. (Still a good strategy today.)

This brings us then, to the next Big Thing; the civilization of Greece. The miracle that was the Greek civilization is said to have begun about 500 BC, although it had numerous antecedents over a wide time span and a wide geographical spread. This civilization was not devised from scratch, but a good deal of it borrowed from all the previous Empires and civilizations and the whole extant human knowledge base.

The first thing to note is that the main contributions of the Greek civilization were not technological advances, but

50

advances of the mind and spirit. The whole Greek experiment began in the midst of a tumultuous world. It was whirling with new horizons, hurly burly 'worldwide' commerce, wars, nearby empires threatening themselves and the Grecian alike, and clouds of uncertainty, if not chaos gathering all round. It is in retrospect the greatest of all Greek tragedies that the height of Greek civilization only lasted 150 years.

The Greek experiment started humbly as a group of scattered villages which became one by one, city-states. Exactly when these city-states in Greece got the idea to be a "polis" is not known but by 800 BC the process was entrenched. It was then deliberately spread around the region by sending colonists around to several already good-sized cities. The individual city-states had from the beginning, a bent toward egalitarianism and also a curiosity that was rare, unprecedented really, and uncharacteristic of all that had come before.

Athens very early on became the soul of the movement but all the other city-states remained always autonomous. They never became consolidated into a Nation, although they had alliances and close trading partners and other relationships. By the middle of the sixth century BC, Athens began to democratically elect their leaders, the first democratic governance anywhere.

But by the turn of the century into 500 BC, all of the Grecian city-states, as well as the whole known world, was menaced by the new Persian empire. By June 480 BC their King Xerxes had invaded all of Greece and was at the gate of Athens. It appeared all hope was gone. But in September 480 one of the most dramatic events in human history occurred, orchestrated by Themistocles. Themistocles deceived King Xerxes into believing that the combined navies of Sparta and Athens (which had been badly damaged), had retreated. Thus, Xerxes sent his navy to destroy it once and for all. But Themistocles had set a trap for it off the island of Salamis and destroyed Xerxes' navy, ending the Persian threat.

This ushered in almost immediately the golden age of classical Greece, with its magnificent art, theater, architecture, science, philosophy and politics. Behind it all, lay a passion for knowledge, truth and beauty which flowered seemingly out of nowhere and still shines around the world. I will not narrate the great events and accomplishments, nor the inglorious premature end, brought about by the usual suspects of conflict and war. The historian Thucydides describes the ultimate tragedy, beginning with Pericles' disastrous Peloponnesian wars, then the Sparta debacle in 431. A hundred years later, Alexander the Great's father, from nearby Macedonia conquered Athens and just two years after that 20 year old Alexander started his conquest of the world. Alexander however, had Aristotle as his childhood tutor and loved the Greek culture. Except for governance, the Greeks and their offshoots around the Mediterranean were left pretty much alone, a major reason so much of it persisted over the next troubled centuries.

Fortunately, the penultimate great civilization was already arising to save virtually the entire Greek legacy. This was the Roman Republic, and then Empire. Their's is one of the most thrilling stories in all human history, but needs not be re-sequenced here. As every visitor to Rome, and Italian schoolkid 'knows', Rome was begun around 700 BC by Romulus, one of the twins sired by Mars and who was raised by wolves. The year may be close enough but Rome was then only a village by the Tiber like most other agricultural sites. As it grew, along with other Italian city-states, local autonomy fell to kings who consolidated the immediate region.

However, around 400 BC the Romans overthrew the kings and set up a republic. It was this Republic which grew into an economic and aggrandizing military superpower, resulting ultimately in hegemony over most of the Mediterranean territory, Europe and mid- East, including Egypt. Suffice to say, over the next few centuries Rome was fully occupied with its growing empire, each new conquest being incorporated more or less into its own republic. It went on, raggedly successfully, if bloodily, until about the time of Julius Caesar's death in the Senate in 44 BC.

The next 50 years after that event saw some of the most interesting and momentous, if not always virtuous, events in all history. Again all of these are largely familiar, perhaps most famously the infatuation of Marc Antony with Julius Caesar's wife and its fateful consequences in 31 BC. In short, the naval victory of Octavio over both Marc Antony's Navy of the East and Caesar's of the West at Actium, marked the end of the Republic of Rome. The new Roman Empire then began, which lastly nearly half a millenium.

We all know about the great buildings and monuments the Romans left to the world, along with its great engineering and administrative skills, many of which our own civilization has incorporated. Actually, from the very beginning of Rome, even before there was anything but a little village, Greek culture had been essentially deeply inculcated whole, although probably most of the indigenous people didn't really know it. Once Rome became the power it was, it not only adopted essentially Greek thinking but actively promulgated and added to its attributes. This, to many, is in fact one of Rome's greatest achievements. Certain it is, that life in Rome and its provinces was materially improved with the most modern roads, houses, orderly governance, etc.

It is also certain that the political life was strong and extremely hectic and filled with world class intrigue and unending wars, both civil and empiric. Unfortunately, government and politics was not adopted from the Greeks, but was their own unique blend of Emperor and Senate, but in which usually only the Emperor counted. In any case, it was rather effective and brought long term order and organization to most of the known world. Her final legal code represents a major human advance. (Western

53

civilization law basically is underlain by the Roman code.) The Roman's sense of order and discipline was probably what made them so good at law and order and in governance overall. They did not make very many great inventions or contributions to science, or religion or art, as they remained always quite pragmatic. They invented a better system of numbers of course, and this was a major factor in their main contribution to subsequent history in – besides law – civil engineering. Large scale public works, roads, bridges, water systems, communication systems, theaters, coliseums and the like are still held in awe by the talent displayed.

Anyway, by the end of the Roman Empire in the sixth century AD, the rest of the world, even though most of it had been conquered at some point by Rome, had learned something, a great deal from it. Everything humanity had learned since the earliest civilizations, the Romans preserved in some form and it remained out there for all to absorb.

After the Fall, when obviously the throbbing heart had gone out, humanity continued to make, very slowly, a little 'progress'. Day to day life in villages and cities saw slowly increasing trading and commercial enterprises and from time to time some local innovation, which somehow wound up in the whole human race's repertoire. The trend line toward greater mental repertoire and awareness, all without any outward physical signs, from the first feeble man all the way through the Neolithic period to the Romans is very clear. This trajectory is also seen going forward, with increasingly blinding speed into the currently geometrically dizzying speed of our own time.

The Middle Ages and Beyond

To briefly and synoptically narrate some of Man's accomplishments post- Rome we see that much of the next 1000 years was mostly an unfortunate time when there was no center and civilization itself regressed in some respects.

When, exactly the Roman empire finally ended has been endlessly discussed and debated ever since really. For our purposes though, we will take 500 BC, after the last gasp Justinian revival, to be the start of what is mostly now called the Middle Ages.

Just after Rome itself had been conquered – a couple of times – things drastically changed throughout the world. Without a Roman army to worry about, her former conquests, enemies and friends sensed opportunities. It was in fact a confused and chaotic period which ensued, sometimes referred to as the Dark Ages, lasting 300- 500 years.

The period seems to have involved on the macro scale mostly a ceaseless ebb and flow of armies, conflict and head-down persevering by the mass of people. Barbarians do seem to represent the age maybe better than any one word. Hordes of uncivilized tribes, without written language, but skilled in armed conflict, with their hereditary leaders bent on conquest and domination were a constant threat. They came from enormous virtually unknown areas, i.e. the whole northern mysterious Europe and the still unexplored central and northern Asia. Names like Huns, Vandals, Goths, Franks, Saxes, Avars and other tribes from these regions fill the pages of history of those times as they either migrated or marched south and west and southeast; almost always not in peace. They still give us a shudder and must have been the worst nightmare in the civilized areas.

Basically, the next four or five hundred years, from the fourth through the ninth centuries AD seemed to be consumed by armies, or sometimes just tribes and re-confederation of tribes from all these regions searching for better or more territories. In some cases, the groups were merely expanding in response to internal pressure. But in most cases, it was the aggressive ambition of kings, warlords and other types leading their people into wars of intrigue, greed, conquest or aggression.

First, it was the Germanic barbaric – uncivilized – peoples who harassed and harried and attacked more or less continuously at various parts of the Roman lands, eventually throughout Italy itself. The Huns from Asia were part and parcel of this process too from the beginning. Attila himself occupied much of Italy and only the Pope (Leo I) at the behest of the helpless Emperor dissuaded him from entering Rome itself, thus sparing – again – the integrity of the magnificent city. (Attila himself died a mysterious death in Italy not long after.) Meanwhile, Vandals from Germany took over North Africa and especially Carthage. The Franks sliced up Germany and also parts of Gaul and Spain as well as further east. Along with the Goths, this set up – constantly warring – sub kingdoms. Eventually, about the only good thing that came out of this chaos was that the 'barbarians' themselves gradually became more Roman and more civilized. Eventually these occupied countries became the stable regions they are today. (We call them Europeans now of course.)

Except maybe for the enigmatic *Beowulf* written in Britain probably sometime in the 8th century, for 500 years there was little notable new literary output. In fact, art, literature and science were all basically static and were it not for the monasteries of the Benedictine monks, established by Popes Benedict and Gregory, much of the prior knowledge of the great Greek and Roman civilizations would also have been lost. These sanctuaries were pretty much the only repository of education too, as all previous systems were catastrophically reduced everywhere. For centuries the monks selflessly and intelligently copied and preserved a great deal of all the great Greek and Roman literature, history, science and art to the eternal gratitude of mankind.

Another parallel development had been occurring too, all throughout the Middle ages, and even before. This great civilizing enterprise which was taking hold and making some great contributions to mankind was nothing less than essentially the replacement of "pagan" polytheistic religion by the current major religions. Even several hundred years before Rome actually crumbled, the Christian religion arose and grew, even in Rome itself, to an amazing degree. Today, there are an estimated 2.1 billion people who belong to some Christian religion. There are also 1.6 billion Muslims. The Eastern religions likewise claim 1.7 billion adherents; mainly Buddhism with its 400 million followers and 900 million Hindus worldwide. Judaism claims a surprisingly low number – – 14 million.

The last of the major modern religions arose in the Middle East with the birth of Mohammed in Mecca about 570 AD. By the time Mohammed died in 632 he had not only invented a new religion – devised from an odd mixture of religious traditions from Jewish, Christian and his own making – but with his three immediately following disciples managed to spread it amazingly rapidly all through Arabia. Islamic religion grew like wildfire for centuries and is still expanding. Not only had they established the religion but shortly raised a mighty army too, based largely on their religion. Obviously the new Islamic peoples made a mark on history right away, as they took on the whole of the existing dominant powers, namely the Persian empire and what was left of the Roman Empire. They quickly conquered Egypt and North Africa and also much of the remnants of the Roman legacy in Europe. The Muslim armies were very energetic from the beginning and would expand their empire to control vast territories, including all around the Mediterranean and Egypt. This pattern would continue, through the Crusades for example, for almost a thousand years.

Unfortunately, in considering the entire Dark Ages very few other inventions or literary or social or other advances were forthcoming for the next few centuries. Just more religious wars or plain wars of aggression or intrigue, and hereditary rulers of masses of feudalized ragged localities getting by. Learning and even the chance to think about it had virtually become impossible and was still husbanded only in the monasteries and a few outposts in Italy, Greece and other parts of the Byzantine empire. The Arabic empires contributed as much as any other group – besides the Christians – in keeping scholarship alive, especially in their continuing interest in astronomy and mathematics and also with their excellent poetry. Around 800 AD their mathematics expertise gave the world the arabic numbers we still use – although it took four more centuries to spread much.

By about 800 AD, the nadir of the dark ages, an interesting social system had been arrived at in Europe especially. This was the ugly estate called feudalism. By this time there had been about half a millennium of decline and constant war, with power struggles at all levels and the population was actually declining. Life became poorer and harder for all but the aristocracy. There was a very basic, truncated social structure, strictly ruled by kings, lords and noble landowners. Everybody else, save for a few priests, tradesmen and soldiers were agriculturists, that is tilling the soil. However, they no longer owned the soil and their acreages shrank to barely subsistence levels after their landlords and above had taken their lions share.

However, by 800 AD, the Frankish king Charlemagne had conquered virtually all of Europe and most of the Mediterranean area. He stands out as the all time most successful king anywhere in the world, counting from over the span of the previous thousands of years. Charlemagne, himself illiterate, struck the first spark of progress in those centuries during which most of the world had spiraled down and down to abject feudalism. During the middle of his reign – still a record holder for longevity – he seemed to have 'gotten the word' and decided to improve the whole tenor of the time. He recognized that things presented a pretty dismal picture – no real laws or central money, or learning or basis for efficient administration, even tax collection and all the rest. So he began to build beautiful churches and other public buildings and – – good Lord! – – decided that there should be an emphasis on learning.

He wanted to establish schools but hardly anybody really knew how that could even be attempted. So he attracted some of the learned men from all over, particularly a man named Alcuin from England. This heroic effort, under the leadership of the two men, was quite successful, albeit on a small scale. At least it kept a spark alive in the ensuing dark. And dark, after Charlemagne's death, did come again – all over! His successors were no better than his predecessors and the world stayed dark, with just a candle here and there still burning, for a couple centuries more.

"There was no hope for him this time."
Opening sentence, James Joyce, "The Dubliners"

But, by the middle of the 900's, the Frankish incessant battling had finally, out of chaos maybe but by no seeming dint of purpose, produced an ultimately useful result, which was that two victorious sides clearly emerged. These would soon become France and Germany. Also, as a casualty of all the wars, most of the barbarian tribes, including lastly the Avars, Norsemen, Maygars and even most of the Asian tribes, had been virtually vanquished. Actually they either melted into their victors' camp or settled down themselves in their home turf, but at least they were neutralized as a constant siege threat.

Part of the reason for all this is that in the old Frankish empire, the next generation war horses were being raised at this time. With these well bred heavy horses fitted with armor, many of the noblemen as well as the kings invented a new form of army, the knights. These small, elite units proved almost invincible and hastened the end of the widespread, endless wars, especially with the last of the Eurasian barbarians, who became assimilated.

Two other seemingly unimportant, little noted progressive inventions were made along the way. The most important was the moldboard plow, which along with the horse collar enabled the heavy European soil to become more productive. Thus, the horse largely replaced the ox as draft animals. The increased production of food was felt right away and the population began a slow rise. It didn't hurt either that it happened that the last of the Kings of the old Frankish realm were both smart and competent. King Otto I (Otto the Great to history) came to power in what would soon become Germany around 950. He immediately set out to consolidate the country once and for all; and in short order he did. Not only did he bring peace and stability to western Europe, but also to Italy, whereupon the Pope proclaimed him Roman Emperor.

Fortuitously, in the western region, the same kind of thing was happening. Hugh Capet, the young son of the last Frankish king there began to assert his right to ascension. With the help of the Secretary to the Archbishop of Rome, Gerbert, who was acknowledged the world's best scholar, they convinced the Pope to anoint Hugh Capet as King of what was thenceforth called France. Both Otto and Hugh Capet had strong sympathies for learning, scholarship and education. They both collaborated, (especially by the tireless efforts of Gerbert) to advance all sorts of non-theological learning; schools, mathematics, astronomy, books, etc. Gerbert not only knew and liked Hugh Capet, but had also once tutored King Otto. Otto later had him named Pope, which facilitated the work. Although this great progress made over 30 or so years was slowed after the deaths of the two kings, it was not lost but merely smoldered for a couple more centuries before it flamed back up again.

OK, we've now traced the earth and its new inheritor to just 1000 years from our own time. We've traced his prehistory and the bulk of his history and we have declared that the world is ours – i.e. western civilization's. I will trace this last

60

thousand- year history only in general outline too, as some people alive now can trace a branch of their kin back to King Otto's time. Just a brief selection of some of the outstanding or interesting events and people will suffice as a gauge of the accelerating speed, scope, complexity and multiplicity of the changes man has had to incorporate into his stone age brain. Physical evolution was not much of a factor, but social and cultural evolution pretty much ruled (as it has ever since).

At the beginning of the year 1000 there were an estimated 35-75 million people on earth, almost 90% of whom were agriculturalists. Actually, the serfs, who farmed and lived on the land didn't own anything except the right to their own produce – – minus the 'taxes' to the nobles and king. For the next 300 years, not much changed in the way of life for the vast multitude of people. The main current, and it turned out to be a very powerful, if invisible one the whole time, really was evolutionary small changes in the routine. Probably the main factors were slow, imperceptible changes in the relation of farmers – serfs initially – to the land and to the towns and also the relationships of town residents to themselves and to the rulers. The trend, though nobody noticed it until centuries later, was for more and more and larger villages and more distributed government. This meant that there were more towns and provinces with a couple of layers of government to somewhat diffuse power of the usual princes or counts.

Whether Crusading or not, the church was the dominant social force in most places during this entire period. Principally the Catholic church because it was so widespread and organized, but also Islam were the unifying forces. While kings, nobles, dukes and such certainly held great power over their respective bailiwicks, sometimes benignly, sometimes cruelly, it was the religious organizations, especially Catholic churches, which held virtually all men and women in Europe under their daily sway. Islam too, was strong on dogmatic adherence to rigid ecclesiastic doctrine and ritual, thus controlling the structure of the core society

61

too. It may be noted that by the end of the High middle ages, the Islamic empire was by far the largest one – second all-time only to the later British.

Also during this time, there were fewer huge empires with an Emperor holding strong sway over the whole territory. This lead to gradually increasing egalitarianism in the general sense that locals, including the farmers who would soon own some of the land, could feel more secure and less manipulated by all . powerful distant monarchs or rulers. Many towns were virtually their own little kingdom.This was also the time of the Great Crusades but except for Genghis Khan and the Mongols, there were no long and bloody wars of empire builders. There were also many human accomplishments, and from a long view, heartening signs of promise for the human race during these centuries, even though they were more or less unnoticeable by the vast majority of humankind, even in Europe. These bright spots were more like occasional meteors flashing up in the dark whilst all around darkness prevailed. Life for most people was still at a pretty low ebb, a lifelong struggle for subsistence; hard, often cruel and seemingly fraught with inscrutable and unexpected disaster from disease, or war, or warlords, etc. etc. From this far rear view, the little sparks of something in the human mind and spirit that we do see gives us a little insight into a little more of human nature. And a little optimism that the longer the whole enterprise goes the more marvelous and surprising it becomes.

Part of this process was the introduction in the towns of guilds, or associations of differentiated merchants, artisans and soldiers. This, in turn lead to somewhat more sophistication in the whole arena of commerce, banking, accounting and tax collection. Inventions and bits of progress were quite clearly increasing, although still very slowly all the way through to the 13th century. Paper was manufactured in Spain and spread rapidly after the year 1100. Windmills, glass windows, the magnetic compass, eyeglasses and wooden moveable type, for example were also invented.

Even before or just after the year 1200, there were other little signs of progress. A library was established by the Muslims in Egypt and the first universities were opened in Bologna and Oxford and Paris and Cambridge. Avicenna wrote his monumental "Book of Healing" and Viking colonists amazed by sailing vast distances to Vinland. Towns continued to rise. For example, by 1200 AD Paris, Venice, Milan and Florence each had over 80,000 people. Cultural development, like the Latin Classics, started to move to cities from the monasteries where it had mostly been for centuries. Dante Alighieri, Tomas Aquinas, Roger Bacon and the Magna Carta signified other great cultural accomplishments. Great buildings, like Notre Dame and many other churches and halls started to be built on a grand scale for the first time in almost a thousand years. Population had probably reached over 400 million by the end of the thirteenth century.

Then came the 1300's. The whole world seemed to take a strange turn, from no perceived source and certainly any conceived reason. One misfortune followed one cataclysm after another – – all without anyone obviously willing it or understanding any of it. (Unless it was God's wrath, as the clerics warned.) If the previous centuries were Dark this must have seemed to most of humanity, at least in all Europe, that the world had descended into some sort of preparation for Hades. First off, the climate changed, to wet and cold. The Baltic Sea froze over completely several times and the Little Ice Age began. Crops were badly affected and locally real hunger stalked. Hardship was the rule. What little political stability had been the norm was shaken by various wars, including more Crusades and toward the middle of the century, the disastrous Hundred Year War between France and England. The Pope, from years of bungling and
corruption was forced from Rome to Avignon, France,

where serious moral and political corruption continued for much of the century. This made the Church, except in local parishes, almost absent as a unifying force. This church dysfunction caused great angst in people, even to the point where the stirrings of the Reformation can be seen. In any event, throughout this century basically no progress was made in any direction, from art to politics.

In 1315-1320, the Great Famine struck most of Europe, due partly to the extremely poor weather and serial crop failures. Large numbers of people starved to death and the average life span fell to unprecedented levels. Then, in 1347, the Black Death struck, apocalyptically, to decimate different cities and nations in waves for the next several years. Within days people were dying everywhere, striking terror everywhere. The populations of cities especially, but also villages and farm families were decimated by the unknown plague and normal life was impossible. Terror and social destruction ruled. The sight of a neighbor coughing terribly ('churchyard cough' in one of Melville's great phrases) sent people, even family, fleeing. The population of England for example, went from 4 million to 2.5 million and throughout Europe it is estimated that almost half of the whole population died in just a few years. From 75 to 150 million people died.

Turning misery and despair into tragedy was the order of the day, for decades. If the Four Horsemen of the Apocalypse ever did course the land, this must have been the time. Most tragically, the effect on the countryside food production was devastating. Land went unplowed, herds went untended, and not only death from disease itself, but hunger and starvation were virtually everywhere for several years until the plague subsided, village by village. Food price inflation was immediate.

An unplanned benefit became visible though after a few years, with huge significance ever since, as the need for more workers to work the fields became acute. Landowners and nobles were forced to pay higher wages to induce people to become agriculturists. This led to better living conditions and incidentally in raising the political status of the rural people, a condition which would become for the most part, permanent. All this time, the stress and turmoil led to a general disaffection with both the Church and government which were seen as helpless and useless. By mid-century the population of the world is estimated at only 350 million. In fact, all these catastrophes, mainly in Europe, became a significant reason that the feudal system was soon doomed, with far reaching effects.

By 1400 AD, the world begins to pretty much turn upside down; more aptly, right side up, giving rise to the Renaissance over the next two centuries. Throughout the 15th century, politically and ecclesiastically, the still feudal structure became more stabilized. People could now turn from suffering outrageous oppression from divine, malign, or mystery forces or diseases, to trying to improve their circumstances. Creative juices seemed to flow up from many spots, as learning or religious freedom were not subversive activities so much as in the past. There were still Inquisitions, stake burnings, mob massacres, palace intrigues, wars and all the rest to be sure, but now, especially at the end of the 1400's, there begins to be also more 'sane' people around to give pushback.

Early in the century, in fact, several Universities, like University of Turin and Leipzig and Prague for example, were started, also the Bethlehem Hospital in London. Agriculture was stable, allowing the population to begin growing rather rapidly again. Population was over 435 million worldwide at the end of the century. By 1434 banking and commerce had blossomed and the Medici banking enterprise in Venice and Florence was flourishing. The first copyright, with all that it portends in law and commerce, was granted not long after in Venice. In mid - century, Gutenberg changed the world with his printing press. Significantly too, many of the giants of all humanity had already been born or soon would be. Botticelli, Jan van Eyck, da Vinci, and Michelangelo and other towering figures were loose in the world.

In 1487 the age of worldwide exploration was ushered in with Bartholomew Diaz discovering the way to the east around the Cape of Good Hope. By 1498 De Gama went all the way from Portugal to India and back. Already Columbus had sailed a couple of times to the New World. Twenty years later Magellan and crew would conclusively prove to an amazed world that one could sail around it. The Cabots and others were already exploring and claiming Canada, as was de Soto inland in the future U.S. Luther shook the world forever in 1517 by posting his famous "95 Theses." Raphael would soon die but the letter "j" would be introduced into the alphabet. I should not neglect to mention though, besides the
incessant wars in England, Asia, Europe, Near and Middle East and Africa, the Spanish Inquisition just to round out the picture! Population of the world was about 570 million by the beginning of the 1600's.

Whatever was in the cultural waters to allow some of these new progressive forces to pop up is an intensely interesting question. In any case, it was a growing force and burst out in spades in the 16th – and also the 17th , 18th , 19th, 20th, 21th and hopefully 22nd Century.

Mid-sixteenth century the Elizabethan era began, which corresponds to the full flowering of the Renaissance with Operas, flush toilets, John Donne, Thomas More, and of course, Shakespeare. Galileo and Kepler are born and Roger Drake explores around the globe. From this point on, we all learned in school the quaint path leading to the Modern period – Western Civilization – with all the famous names, dates, and wars and wars and exploration etc. In the 17th century we learned of the colonization of North and South America. We also learned the actions of the various kings and potentates of Europe, the Near and Mid East and their

66

wars. We were also bequeathed the great art and literature that are basically ours today.

We learned that the world beginning back then acquired electricity, telescopes, microscopes, Galileo, Kepler, Newtons Laws, Boyle, Hooke, van Leeuwenhoek, Pascal, Descartes, Hume, Locke; in a word, science, as well as a "'real" society. Everything that happened was written down. And what was written down was written upon, and expounded upon, and argued about ever since so we can trace not only the physical changes but the mental and cultural evolution that sticks to the history. We retrospectively all see at once all the great progressive changes that arose, slowly at first, then faster and faster right up to the latest 'Cloud'. What kind of mental itch or ferment must somehow have been involved!

How much did the basic social structure and the psychological structure, deep down, really change since the Dark Ages? I am incompetent to even begin to analyze these at any satisfactory level, but I do think that deep study of social archeology and psychological archeology would be rewarding. One example might be the saga of Galileo. Galileo was born in 1564 and died in 1642. This was one of the greatest scientists who ever lived, not counting Aristotle and Descartes who, between them and Francis Bacon, virtually invented the scientific approach. Basically, Galileo confirmed and proved, what Copernicus, and later Kepler, had hypothesized, i.e. that the earth revolved around the sun, not vice versa. Galileo used, almost for the first time, *observation and experiment,* to show the sun's and the earth's orbits. For this, he was brought to trial in 1633 by the church tribunal – actually the Pope. He was sentenced to life in prison for heresy ".... for holding an opinion declared contrary to Holy Scripture." Could have been worse; he was allowed to remain under loose and fairly comfortable house arrest until he died. (Later on, the great Rene Descartes too, would be hounded to death by similar, even worse persecution.)

67

I might try to sculpt here an interesting fact out of Man's whole history over these past 5000 bloody years of civilizations: his greatest misfortunes, misery and unhappiness arose, not out of himself, from some evil inside him, but from clashes from outside, from his fellow man writ large. Big Trouble most often came from his organizations which surrounded him. He couldn't help it of course. Wherever he happened to be born, whether in a yurt in Mongolia, or a hut along the Rhine, or even Paris or Italy, he came already bound to a Principality of some type, and a Prince of some kind, whether a King, a noble, a religious, or other oppressor. His 'Organization(s)' are what enslaved or oppressed him. It wasn't usually his neighboring farmer or the next door villager who caused most of his grief. What few seemed to grasp until late in the 18th century (say about 1776?) was that his organizations are actually voluntary. The people themselves could (should) decide them collectively

and furthermore that they are responsible for continuous further modifications as decided upon from time to time. Better Organizations were needed then as they are now; fortunately they are still in the process of being, very slowly, invented and reformatted (cultural evolution) by societies themselves.

At any rate, by 1900, which year we now think so quaint and long ago, our literature and art galleries were already filled with da Vinci's, and Rembrandt's and Manet's and Monet's, of which, again, we are the inheritors. The grandest architectural structures and machines, which are unsurpassed, some from over 3000 years ago, we have inherited. The history of just the last 100 years makes our heads spin off. But this too, is different not in kind, but only in degree from our previous 10,000. Are we better off? I think so, but it is an extremely important question, to which we will return later. Anyway, can we agree that it is a part of the continuum from 70,000 or more years ago?

Of all these people then, starting even from *Ardipithecus*

the ape, we are the inheritors. Our bodies, our selves, our world. By the end of the 1600's, America had three quarter million people, counting Indians. The world was up to 900,000,000 by the end of the 1700's and man had crawled all around the world. Within 250 more years our super successful species would cover practically every square foot and OWN the world. No other species ever has or likely ever will completely rule everywhere at once. By 1800 our western civilization was poised to do anything, go anywhere, invent whatever to follow some innate urge, born of curiosity perhaps at the core, to explore, expand, control and understand all the world that he had inherited.

I will conclude this survey of world prehistory and also history and in the next chapter look into how the past has influenced us modern humans embedded in the latest model of global civilization. How did a human with this kind of history make us the way we are today? And with the mind and culture that we have today? Obviously, both our bodies and mental and culture characteristics are the products of that history. I start with a closer look at some of the processes of physical and cultural evolution that are now thought to have been the main mechanisms. Hopefully that will help lead us to fruitful further exploration.

CHAPTER III: EVOLUTION AND HUMAN NATURE

The question is "what is the nature of man and how do we explain his incredible mental, social, and cultural features and how they came to be?" Our answer in a word is still--evolution. Our short and truncated treatment here has not been original but simply standard modern evolutionary thought as found in numerous books and papers in the normal press, starting with Darwin. Humans arose in the first place in accord with the standard notion of biological evolution by natural selection just like any other member of the biosphere. His special nature of advanced awareness and Mind also evolved by natural selection, mostly by gene-culture co-evolution over the last million years or so. The overarching idea is usually stated somewhat like this: early man already came equipped with virtually all of the physical brain apparatus as it is now; the observed mental progression over the next million or so years was the result of natural selection of various traits by both (a) Individual and (b) Group (social) evolution.

In addition to a vast scientific literature, many books (especially E.O. Wilson's recent books "The Social Conquest of Earth", 2012, "The Meaning of Human Existence", 2014; Damasio, 2010, Gazzaninga, 2018 and Dennett, 2017) give especially good accounts of the current thinking about the origin of Mind. These are still- evolving, evolutionary arguments about the roles of the genetic physical underpinnings and the pure cultural evolution leading to the mind. Their focus, however, was on explaining how we came to have our extraordinary mind in the first place. Our focus here has been on simply characterizing what kind of Mind and species we actually wound up with (and which we are stuck with).

Going back then, to our previous survey of the origin of the human species we see that early man (starting at least with *Homo erectus ?)* already had developed a mind and also culture. He lived in tribes and as all anthropological and also animal studies show, group and social living is a challenge. Sociality is not rare in animals, but highly complex and organized sociality – defined as eusociality – is almost vanishingly rare. It occurs in only a few insects (mostly bees and ants) and, except for man, one other mammal.

In all the other social animals, too, like wolves, primates, cows and a host of other familiar examples, there are many social instincts that comes inbred. That is to say, the brains of these animals have a genetically determined social propensity, which arose solely because it produced a selective survival advantage in the particular environment. At the least, this involved gathering with their own kind and not generally indiscriminately killing their own.

The way the brain works unconsciously tends naturally to this urge to form in small groups; in the case of humans, a band or tribe of around 3-8 dozen or so individuals. From the first tribe on, the survival of the tribe was primary, and thus the characteristics that enabled better survival of the tribe became rich, successful targets of natural selection. Wilson summarizes a great deal of recent anthropological research : "A basic element of human nature is that people feel compelled to belong to groups, and, having joined, consider them superior to competing groups..... . .. an iron rule exists in genetic social evolution. It is that selfish individuals beat altruistic individuals, while groups of altruists beat groups of selfish individuals."

What were some of those social characteristics that evolved? We are in the infant stages of sorting them out, but the main ones are well accepted. First off, the human brain comes spring loaded with social urges – needs, even – to form cooperative groups. Prehistoric man as noted earlier, took a first step toward social development by the absolutely human

71

step of gathering together at campsites, the Wilsonian homologue of the Nest. They gathered there daily, foraged for food and brought it back to share, raised their young there, and defended their territory. From the first, it was a social, cooperative enterprise.

The adoption of a diet containing a lot of meat, with all its physical, psychic and social challenges was a significant factor in producing both individual and group characteristics of humans. Finely gauging the intentions of others, enabling growth in the ability to gain trust and also manage rivals was/is an important part of humanness. Social intelligence was therefore always at a high premium. It pays to be socially smart.

Further, the group members inevitably compete with one another for status, mates, etc. The pathways to eusociality Wilson says "was charted by a contest between selection based on the relative success of individuals within groups versus relative success among groups. The strategies of this game were written as a complicated mix of closely calibrated altruism, cooperation, competition, domination, aggression, reciprocity, defection, and deceit." All this required fine-graded evaluation among group members, leading to increases in the trait of intelligence as well as all the other aspects of the mind.

Quoting Wilson again " To play the game the human way, it was necessary for the evolving population to acquire an ever higher level of intelligence, not to mention morality, emotions and all the rest. Thus was born the human condition, selfish at one time, selfless at another, the two impulses often conflicted." The human condition is, therefore, an endemic turmoil rooted in the evolutionary processes that created us. Much of culture has arisen from the inevitable clash of individual selection and group selection traits, i.e. between our two, or more, types of brains.

Genes have been termed "selfish genes" by Dawkins and others. In a sense it is true, that all genes are selfish and in fact, it is the individual gene, not individuals that is *the* unit of evolution. (This is a difficult concept for even some scientists to fully grasp; see Wilson for a clear argument.) Truly though, what are more commonly called selfish genes today are those 'individual' genes which lead usually to traits such as self -interest, self- aggrandizement, dominance, selfishness, deceit, extremely strong loyalties to a restricted in-group, implacable status seeking, and the like. They are the me-first type of traits and are designed to ensure that the "I" survive and thrive. Humans clearly have a goodly number of these type selfish genes, probably a plurality. These are strongly engrained in our brains and are a large part of our human nature. Those parts of the brain that mediate these various behaviors are hardwired in the sense that we all have these modules, all are primed to react in their own rather constrained way. They are also ineradicable, just as is our other "good" brain parts.

The countervailing force to those individual, selfish genes are the 'good' genes. These give us those traits that are not primarily selfish. Ones that tend to make the person altruistic, given to cooperation, conscientiousness, high empathy and similar that relate to the good of the tribe as a whole. These are the genes preserved from group selection which favor cooperative behavior, empathy, love, more sensitivity to the whole group and to give aid even if the cost to self is rather high. These too are strongly engrained – inherited – in the human brain by virtue of its evolutionary path. All of these things are the raw materials of social evolution. They arose by selection for genes for these traits more so than others in their own or other tribes.

Philosophers have always, correctly, noted our dual natures and we now know, partly, why. (I speak here not of the mind/body duality, which I reject, just the common two types of patterns of behaviors noted above. Also, I cannot emphasize too strongly the absolute wrongheadedness of the whole, hopefully now discarded, idea of social Darwinism

73

which grotesquely tries to apply the laws of genetics to the results of social intercourse and cultural mores.)

Thus, if the tribe was pre-eminent in the ever more complicated human evolution – and it was – then selection of the whole tribe by forces of the particular environment must have often overridden that of any one, or many, individuals in the group. If the whole tribe goes under, then what good is it if a couple of their individual members happened to have had a very felicitous genetic makeup that might be considered desirable to have if all else were equal. In a tribe, the whole really is greater than the sum of its parts. Group, or cultural selection, therefore, is now considered the major force that made us what we are. Those genes which contributed traits advantageous for survival of the tribe were acted upon by natural selection and some were kept.

So, again, we see that from beginning to end, from the first bipedal, savannah- bound, completely ground living human onward, we humans were tribal and were under very considerable pressure from the environment for survival. They lived or died as tribes – in numbers there is strength – so the characteristics that facilitated tribal survival were critical for a long time and so, behold, the physical brains we have today. Hardwired into them are the various spring-loaded modules which, not automatically but most naturally, lead to final decisions to behave in certain ways under certain environmental situations. It is not, as I emphasize elsewhere, that our brains are totally, or even to a major extent comprised of hardwired circuits to automatically result in certain behaviors. These latter kinds of circuits would include reflexes, of which humans have only a few, like the sneeze reflex.

But we do have numerous instincts, most of which are subconscious. These then often trigger off and play into our more or less final complex behaviors. These are extremely powerful circuits and structures and affect our higher behaviors very powerfully all the time. Instincts in humans

74

are not as obvious or conspicuous like a peacock's instinctive tail plumage display for example, but many are quite complex, like sex instincts, or reaction to fearful objects. Even a thing like greed is considered an instinct in animals and man. Fear, anger, rage, etc. are also familiar instincts. (Gazzaniga's book, in fact is titled "Consciousness is An Instinct".) Without these hardwired mental modules activated by environmental stimuli, there would be no species of course.

To broadly sum up our very wide snapshot of our tribal brain and mind, we see that it is partly the end result of individual competition for survival and reproduction among members of the same group. This selection shaped many traits and instincts in individual members that are fundamentally selfish, with pronounced tendencies to aggressiveness, dominance, hierarchies and similar familiar traits.

In contrast, group selected traits are the result of competition between tribes or societies, through both direct conflict and differential competence in exploiting the environment. It became, in fact, the dominant factor in shaping *Homo sapiens*. Group selection results in a different set of traits and instincts, those that tend to favor individuals who are altruistic, empathic and cooperative toward one another (but not necessarily toward members of other groups – a very important point). So we have these two competing, pretty much opposite genetic traits in our brains. Thus, since neither the selfish nor the social group of genes are going away anytime soon we are going to have to learn to deal with it.

Any reading of history shows (a sample of which we have summarized above in our trek through civilizations) how the brain actions of princes, popes, city-states, nations, empires, or territories led sometimes to comical, sometimes good, more often, tragic actions for people. Actually, if examined closely these illustrate the workings out of the two brains.

75

As modern psychology amply documents, humans have a dazzling array of inbuilt bias modules; self bias, group bias, optical and cognitive and other illusions, halo effects etc.

 (See Kahneman, 2011 for a hundred others, many of them surprising and/or hilarious that operate pretty much full time in all people.)

 And science is beginning to show the multiple anatomical or functional parts of the brain in which these conflicting traits reside. Thus, any time a person is presented with some choice (which is almost continuously) there are the fears, emotions, thoughts and instincts which come rushing in simultaneously from the two sides of the brain. Broadly speaking, these come from both (a) the individual, selfish side, urging through its specific brain modules the me- first, individual choice and also (b) from the other, social side, urging one to choose to do more or less the opposite, group oriented, behaviors. Hence our inherently dual nature, our two minds.

 Our inborn brains, much of it subconscious, operating all the time makes us at times absolute slaves to one or another of our emotions. At other times we are fully engaged in fully conscious deliberate high- level intellectualizing. As we will look at briefly in the next chapter, a great deal of current biology, medicine, psychology, and neurophysiology is devoted to identifying not only the particular behaviors but also the neural structures and mechanisms by which these might be mostly effected.

 Simple tribal behaviors served early mankind well, but when we get to more complex living arrangements, like villages, settlements, towns, and finally cities and states, it seems to be a whole new ballgame, even if we are stuck with the brain structures for the simpler situations. Even before our civilizations began, people were presented with more or less constant contacts with people. In fact, almost from the beginning, people were the major factor in their environment, thus offering more or less constant necessity for interaction and fast reading of the intentions of other members, etc; in effect creating their own environment. Information was/is the new medium of value, and information processing was/is the most important \environmental variable. (See Dennett, 2017, for an up-to-date, delightful and illuminating discussion.)

This conflicted brain and pychology has been, and is, the crux of many of the devils afflicting modern man and his civilization. This tension creates most of our moral dilemmas. The concept of morality too, developed within this struggle, from rats onward and is still developing. Unfortunately there is no taxonomy or even any kind of agreed upon philosophic code for this problem with our minds which we are stuck with now. Johnathan Haidt in his latest book "The Righteous Mind" gives an excellent purview of what he calls the Moral Foundations which seem to drive most human psychology and life. Our tribes now are often nebulous, ideological and even only theoretical long range connections. Fortunately (or perhaps unfortunately) we actually have no physical tribe around to give constant feedback, guidance or succor like in the long ago. We are, in fact, alone and 'flying blind by the seat of our pants' so to speak. However, it is still an innate drive for every person to 'belong'. Those groups to which each finally does choose to belong, often subconsciously, in the end then becomes his tribe(s).

Calling the other (political) tribe racist or homophobic and the like is of little help, although it may be literally true, as our subconscious brain sees things. Tribal hatred is obviously of little use in today's melded culture, but until we learn, with the help of science, education, etc, to condition our brains during childhood, youth and throughout life, the full-fledged whirrings of our primitive brains will too often prevail. There are grounds for hope.

CHAPTER IV: INSIDE MODERN MAN: NEUROBIOLOGY, MORALITY, AND MORE

Now that we've traced modern man in all his glory, we will investigate a bit more closely some of the neurobiological and cultural features of mind. This is a difficult, impossible really, task as it requires clearly delineating between brain and mind, as well as thought, imagination, consciousness and other slippery concepts. Therefore, I'll have to cheat quite a lot and throughout just use these terms as they are generally used in standard workaday currency. My aim is limited, and especially on how the physical brain might transduce the interactions between an individual and the environment, i.e. behavior. For this, we'll just do a brief overview and give a few samples of the science of the brain and mind.

First, a very simplified overview of some of the major structures of the brain that mediate the various aspects of the mind. Throughout evolution in the vertebrates the basic plan for the brain has been for a tripartite brain, eg. 1) a Hindbrain (the most primitive, comprising the medulla, pons, cerebellum etc. which are used mainly in the vegetative, subconscious control of the somatic functions of the body); 2) the Midbrain (the so-called reptilian brain); and 3) the Forebrain. The forebrain, the newest, largest and fanciest part of the brain, is the dominant one in all higher mammals. In our case this (mostly cortex) is hugely engorged and infolded and occupies over 75% of the total brain.

Figure 2 below shows some of the basic features of the human brain (cut in two to illustrate some of the major parts of the brain). The most obvious feature is the endlessly folded cortex of the cerebrum (Forebrain) which is crucial for higher functioning, including cognition, learning, mentation, art, civilization, etc. However, the majority of our brain's daily functioning is not primarily the result of this brain, but instead it's the midbrain, the reptilian one, along with the Hindbrain, as explained a bit later.

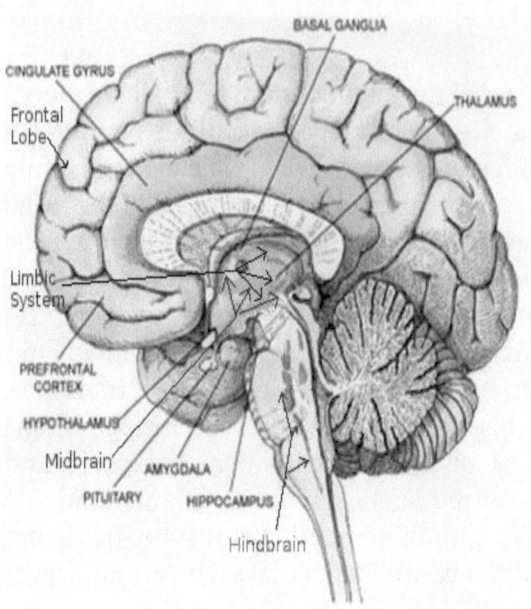

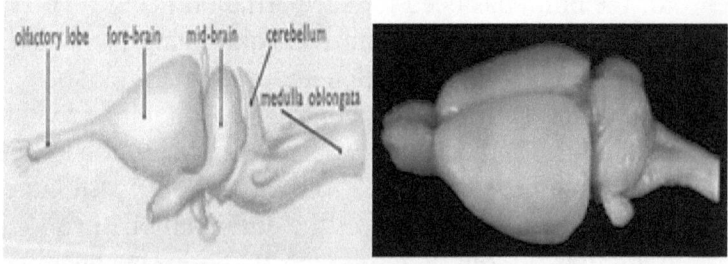

Figure 2. Diagram of a Sagittal (mid) section of human brain showing a small sampling of major functional areas. The photos above are photos showing the more primitive tripartite structure, of a rat brain (right) and a lizard brain (left).

The photos of a rat brain and a lizard brain are shown mainly to indicate that (a) the lizard brain is mostly mid-brain and hindbrain, and (b) even the cortex of the rat brain is smooth. This indicates that their small and unfolded cortex, the seat of the highest brain activity, is minimal and largely undeveloped. Also, just as a couple of examples, a few of the special areas of the human brain are labeled, like some of the frontal lobe's further functionally distinct areas, such as prefrontal cortex, cingulate sulcus, etc. This dominate frontal lobe, of course, is what is probably meant in ordinary conversation by "brain" or head, as it is the critical part of our highest intellectual and mental achievements.

Some of the general parts of the midbrain are also labelled, especially the limbic system, the amygdala and hypothalamus. We will look further into some specific functions of a couple of these special areas in a bit, but for now the main point is that in general each module – i.e., groups of neurons, dendrites, glia and axons, etc. – reacts to and generates or controls particular, rather specific things. The midbrain structures, for example, may be crudely described as the seat or origin of most instincts and emotions – i.e. feelings, thoughts and states of mind we casually call 'passions', 'gut', or 'heart'.

In general, while the various areas are more or less specialized, it is the interconnections and neural traffic patterns between cortical nuclei, other scattered modules and lower brain structures, like the "reptilian" limbic system, hypothalamus and all the rest, that may be in the final analysis the secret to its ultimate functioning. There are around 80 billion neurons and each of these have as many as 10,000 connections to god knows where, so the number of potential patterns of neural activity overall is for all practical purposes almost infinite. In addition, the frequency of action potentials – nerve impulses – in any neuron also carries biologically significant information. This ranges from zero to 100 per second and raises the complexity to another whole new level.

So, this is where the final complexity may come from, somewhat like in supercomputers – and ten thousand times greater besides. It needs to be noted too, that our brains are constantly changing a great many of these connections, both as we develop, and as memories accumulate; even when we're sick, or suffer trauma, or learn, etc. It is a goal of neurobiology and medicine and science generally to eventually determine the main firing and wiring paths of each structure in the brain, and especially their individual neuronal *connection*s.

This latter is called the "connectome" – analogous to the genome, proteome and metabolome. Several collaborative efforts have been started, such as the Human Connectome Project, the European Brain Project and the U.S.'s new major Brain Research Project to actually do it. Already many individual connections have been mapped. Many new Brain Atlas's too, including one specifically devoted to genetic analyses have been well started in just the last couple years and these continue to proliferate. Scientists at the Allen Institute for Brain Science have, for example very recently reported the first initial 3D wiring map of the mouse brain, which has 'only' 80 million neurons.

"I like learning stuff. Keeps my brain from nodding off, or worse yet, wandering off" unknown blogger 2014.

Actually, the entire connectome has already been completely mapped for one organism, the scientific workhorse, the roundworm called *Caenorhabditis elegans*. This worm has only 380 neurons to trace, but a look at its small connectome is frighteningly complex. Incidentally, this worm has also had its entire genome mapped so it will be interesting as scientists marry particular genes to particular neurons with their unique function. Of course, even after all these connections and wiring and firing patterns have been mapped out, we will still be a very long ways away from understanding just where the Mind, along with all its ineffable qualities, comes from.

I will make no attempt at any in- depth survey of the vast literature of the brain or psychology, just a little of the flavor of some of the recent neurobiological and neuropsychological findings. My aim is limited to try to illustrate some of the physical brain's features and also to note that these are the underpinnings of what evolution actually acted upon to produce man in the first place. This whole field is leading to what will be a revolution in our understanding of the brain and mind. (For lucid and excellent details of the evolution and functioning of brain and Mind, see, again, especially Damasio, 2010, Gazzaniga, 2018 and Dennett, 2017.) It is well to keep in mind that the brain structures, and the functions built into these, have been established in much the same shape 200,000 years at least, and will not be much changed for centuries, or millennia.

All this, however, will not help us much when we consider the questions of what a mind is for, or aesthetic or moral questions. The only truly big changes in Man really are *immaterial*, i.e., mental and cultural. This is where we will rely on art, philosophy, educators, humanists and thinkers of all types to help make sense of the world.

We will just look at a few examples of the studies designed to ferret out details of how the brain works. It has been found, for example, that a little part of the brain, the anterior insular cortex, is necessary for high levels of self- awareness and for those feelings which "give life meaning". If this area is injured, there is lack of self- awareness and of certain 'feelings'. Also, a new cell type, called VEN cells, has recently been discovered in humans, exclusively located also in this anterior insular area. (The same type of unique cells has recently been found in this area in monkeys too; Evrard et al, 2012.)

Similarly, in the very anterior part of the neocortex there are little areas where sensations of emotions, like love, hate, resentment, self- confidence and embarrassment are integrated. In fact, much of the Frontal, Dorsolateral Prefrontal and nearby areas are *the* main sites of the highest human capacity. If certain points within these areas are damaged, very odd, specific changes result; the person

82

becomes apathetic and has an inability to tell what feelings he or even other people might logically be experiencing. Other functional and mental capacities are untouched. In some who have suffered brain injury in the frontal cortex area, nearly complete lack of empathy and human feelings are the main clinical findings, similar to a psychopath. (BTW, I am continually baffled by the lack of attention given to this group of humans by both science and the public, as they represent almost a distinctly different kind of social human. We will come back to this a little later on.)

Other studies, such as one by Shamey-Tsoory et al in 2011 titled "The Neural Basis For Empathy" also gives a good taste for the type of research that excites neurobiologists. She used neuroimaging techniques in normal and also brain - injured people and found locations in the inferior frontal gyrus and inferior parietal lobule which mediated the recognition and analysis of emotions. Similarly, the ventromedial prefontal cortex (VMPFC), the temporo-parietal junction, and also the Medial Temporal lobe are central for self -reflection and autobiographical memory as well as a general emotional alarm bell. Dozens of these types of interesting fMRI (functional MRI) and other brain imaging and stimulation studies are reported every month.

The dorsolateral prefrontal cortex (DLPFC) is an extremely important, large, well known area which is essential for the highest level cognition. It is not the only such area, but the main one. As the evolutionary psychologist Joshua Greene ("Moral Tribes", 2013) and others have shown, this area is the one most involved in tasks which underlay delayed high level decisions on utilitarian type of tasks, e.g. complex cooperative social decisions.

Joshua Greene described studies using fMRI in subjects which were given a mental task to perform, such as a standard Moral Dilemma problem. He characterizes the prefrontal cortex as a whole as the main site of our "manual mode camera". He uses the metaphor of a camera which has both an innate, automatic, pretty much pre-programmed mode, and also a 'manual' mode, which must be dialed in by conscious thought to flexibly cover an infinitude of situations. The automatic mode he says is point and shoot in which you get the image which is built in by the "manufacturer's" inherent generic features. In our case, this mode resides in numerous areas throughout the brain, but particularly the amygdala, limbic system and VMPFC.

The manual mode, on the other hand, bypasses or modifies all those automatic features. But it also means you have to decide just what parameters you want on your picture. This is the sophisticated thinking mode, involving most critically large parts of the cortex. Greene says the manual mode largely corresponds to the critical function of the prefrontal cortex. It is the thoughts and behaviors arising from activation of this area that create the reasoned, logical, highly intelligent thoughts and features of humans. These underlay the civilized aspects of group behavior, i.e., society. This is the mode which I will call upon later as the best way to move forward, culturally, socially and governmentally.

The automatic mode, then, often is the gut reaction, the emotional side of the brain. This is the most ingrained, frequent, sometimes instinctual side of human nature. Again, these type reactions are coming from various parts of the brain, but especially the midbrain – reptilian – areas such as the amygdala, ventral medial cortex, limbicsystem and others. Fear, anger, disgust, sex, aggression, selfishness and similar emotions are their dominant impulses. However, sometimes pleasure, love, even ecstasy and similar profound feelings often originate there too.

Basically then, in general, we have two, often antagonistic behavior centers;

   1) the 'heart' (amygdala, ventromedial prefrontal cortex, limbic system etc.;
     and
   2) the 'head' – dorsal prefrontal cortex mostly, although a great number of other scattered modules are also involved.

Creating the eternal conflict! Again, we all have these, they operate more or less all the time, and are basic salient biological facts of human nature which we are stuck with.

Pooh" said Rabbit kindly "'you haven't any brain."
"I know" said Pooh humbly

Countless other studies in man and animals on the activity of various brain areas that underlie much of human behavior are eventually going to have to be taken into account. Another example of such studies involve the so-called "pleasure centers" of the brain, i.e. specific areas of nuclei in the cortex especially. (These nuclei are found in all mammals actually.) In certain of these specific brain clusters, the neurons secrete the chemical *dopamine* as their neurotransmitter. Witten et al, 2011, implanted molecules which are sensitive to light into cells of one such nucleus and studied the responses when these were stimulated by tiny flashes of light. The nerve cell responded by sending out nerve impulses which, when they reached the endings, released clouds of dopamine. Significantly, these axons projected mostly to the prefrontal cortex (Dopamine is known to produce sensations of pleasure of some kind.) These rats would then trigger the light pulses themselves and what was seen was a frenzy of lever pushing, for the presumed pleasure high, until total exhaustion! Not a pretty sight.

Similarly, oxytocinergic neurons – neurotransmitter is oxytocin – reside in certain other cortical brain sites. Oxytocin is also well known to produce feelings of pleasure, calmness, lovingness and the like. Oxytocin, even when only a small amount is sprayed into the nose, for example, increases empathy feelings most especially. These effects of dopamine, oxytocin and similar natural brain hormones are widespread in both animals and people.

Somewhat similar studies show for instance, that when an experimental subject administers altruistic punishment to another subject the anterior insula lights up bilaterally. This area of the brain is also known to be activated by pain, anger, and disgust. (Altruistic punishment is disapproval of free-loading or other social transgressions in a group. It is a universal natural response in human groups.) Some other interesting studies involve the so-called ' mirror' neurons'. These are based on the observations that animals almost universally have the same brain response to the same stimuli. In monkeys, watching a partner perform a simple task, picking up a treat for example, causes nearly identical increased neuronal activity within the same areas of the brain in both partners. ("Monkey see, monkey do".)

Similar experiments in humans show the same thing. For example, Mukamel and colleagues measured the level of neural activity from the relevant motor as well as sensory areas of the medial and temporal cortex and also some hippocampal sites, of human subjects while they either performed or watched a simple task similar to that described above in monkeys. As in monkeys, activity in these specific areas was virtually identical in either the doer or watcher. These studies illustrate how imitation, or the ubiquitous drive for social conformity might be transduced.

Recently a group led by Dr. Thomas Sahlapter at the University of Freiburg reported a deeply satisfying real world success of neurobiology. They succeeded in largely curing (a word not often seen in scientific reports) long-standing, severe depression in a group of patients. They did it simply (!) by placing a tiny electrode deep in one of the midbrain structures (one of the 'pleasure centers') and periodically providing sophisticated mild stimulation. Over half of their patients reported a long-standing relief from their behavioral depression. All the rest had much reduced symptoms and could lead a normal life. This is a pretty profound response of the whole 'self' to a small, precise environmental input, showing again the essential connection of a person's mind to his environment, not only his immediate environment but his whole humanoid history too.

Again, all of the studies outlined here are but a tiny sample of the large amount of good science being directed to the brain and the behavioral output. A little sample of the dizzyingly extreme complexity and magnitude of the task may be had in a recent (February, 2014), National Geographic pictorial essay called "Secrets of the Brain". But for now, note that this brief overview is merely a beginning peek into the brain (and a very, very cursory, oversimplified and generalized view of how our brain works). These and thousands of other studies should sooner or later illuminate how our highest brain and mental accomplishments might be transduced. The connectome concept mentioned earlier should also allow development of a deeper understanding of our true nature. The connectome recall, is a real, physical, biological structure, consisting of the entire tangled mass of the billions of neurons and connections of the entire brain – basically all the wires plus the wiring paths. These ultimately result in all our thoughts, ideas, behaviors and all the rest.

I should note that the connectome is not only variable but also slightly different in every single individual (as are most elements of the brain). It has long been known that neuronal connections change within an individual throughout his own lifetime, as in learning for example. This variability and inherent plasticity of function, in combination with the almost infinitely complex wiring substrate itself, therefore leads to essentially limitless function. The general feeling of neuroscientists is that by knowing the wiring diagram and also the electrical nerve impulse traffic, a better concept of how the mind might work could be gained. But again, the complexity of the subject and the depth of our present state of ignorance is so vast that few solid guides are possible, as made so abundantly and brilliantly clear by Robert Sapolsky in his new book ("Behave", 2017).

For the past several years scientists have been successfully measuring some of these electricity "clouds" in the brains of people and animals using brain scans of various sorts, like

EEG, fMRI, blood flow studies, and the like. The important fact here is that these studies (see Petri, et al, 2014; Roseman, et al, 2014 as recent examples) have clearly shown repeatable patterns and correlation between certain mental states or particular behavior and the clouds of electricity passing through one module after another. All these leads to the general concept that this nerve activity running throughout a fixed structure – the connectome grid – may be a major proximal cause of all behavior.

Other types of studies help make more clear the fundamental connection between neurobiology and behavior. The long ago demonstration of the actions of surgical anesthetics, which basically disconnects a person from his consciousness, may be the most dramatic evidence for a somato-psychological cause-and-effect. Recently a study at George Washington University by Koubessi et al, showed that the mechanism for this is intricate and complicated, but is related to the complex mechanisms mentioned above. They demonstrated that stimulation of a little area in the subcortical center of the brain, the claustrum, rendered an epilepsy patient immediately and totally unconscious. Removal of stimulation resulted in immediate and total recovery of normal consciousness. It left her with no recollection or effects of ever having lost consciousness. Stimulation of sites 5 mm away had no effect at all, illustrating again, the fine gradations of structure and function with the various brain structures.

A few recent studies of the effects of various psychedelic drugs on these kinds of physiological brain activities also makes a powerful case. For instance, studies of psilocybin – the hallucinogen of mushroom fame – show that the brain activity, and even more significantly, the mental activity of "drugged" people is often the same as in undrugged people under certain extraordinary and unusual situations. That is, the drug induces strong transcendent type feelings; for example, overwhelming feelings of universal love and connectedness, inner peace, Oneness, seeing the face of God

88

and so on. These reported ineffable feelings are pretty much of the same type and power that people like mystics and religious and others who practice "mindfulness" etc. also routinely report. In both cases, these feelings are real, these feelings are powerful, often described as the most profound experience of their life, and they are often of immense value to people (see the plain- titled paper by Griffiths et al, "Psilocybin Can Occasion Mystical-type Experiences Having Substantial and Sustained Personal Meaning and Spiritual Significance", 2006). Of course, psilocybin, as well as many similar acting drugs, e.g. LSD, MDMA and others, unfortunately, quite often also produces less salutary effects, like panic, paranoia, confusion or worse.

These studies on the neurobiology of mysticism, using psilocybin in particular, seems to strongly demonstrate the dependence of the feelings and emotions on the kind of biological substrates we have been talking about. Psilocybin apparently works by somehow affecting some of the normal control systems for neural coordination pathways – i.e affecting the electrical patterning along the 'grid'. Specifically Carhart-Harris et al (2012) found in fMRI studies that activity was reduced by psilocybin in what is called the "default-mode network". This poorly known but powerful and probably pivotal network is comprised of an aggregation of brain modules scattered around various brain areas. It is believed to critically link parts of the higher cortex to deeper, older structures of the brain, including the hippocampus and limbic system. Carhart-Harris (and others too, like Damasio) refers to it as the "orchestra conductor", who is charged with managing and holding together major parts of the whole system.

But the main point I want to make here is that these feelings, and all others too, according to the 'Science line', are natural and come from within, at least partially knowable, physical substrates. It is well known that deep and long lasting religious and spiritual feelings ('truths' even) are known to be engendered simply by damage or stimulation to, for example, the right parietal lobe. People having near-death experiences often report similar phenomena of seemingly ecclesiastical or 'beyond human' ineffably strong feelings. Indian warriors on a Vision Quest probably relied upon similar mechanisms. Again, this is not to say that these feelings, insights and the like are not real or of great value to that person. Only that the assertion that they are from supernatural sources is not warranted either. Poetry, mystic sayings, holy books and the like are not, usually, just another way of publicly and reliably 'knowing'.

I do not disagree that these other venues may provide a pathway to knowing things in one's own mind; it's just that these necessarily are not objective, public, knowledge. We should not mix them up willy- nilly, especially in the larger public arena. They can however, like intuition or insights be very valuable additions to the mix and may often be the best source to consult in personal final decisions or choices. But when it comes to society, politics and government, then philosophy or mysticism or poetry alone will not go far. It must be married with Organization, and all the scientific and practical stuff that that implies. Similarly, neither could we come to understand ethics, or politics, or art through a lifetime of studying particle physics, even though these 'particles' are what everybody and the universe too, is made of. (Later on, I'll discuss some of the latest new science - fiction-like stuff and the dangers ahead which could result from science.)

Please note that I am not going to argue that once all the neural structure, function, connections, etc., have all been scientifically identified, tagged, numbered and tamed, will we have solved everything. We will have advanced our understanding, and hopefully even some of the language we use. The technology will revolutionize the way we view the world, but the whole other part of humankind will still be unsolved. The mind, morals, society, politics, philosophy, and arts will still be beyond our ken and have to be unendingly sorted out by other, still unexplored, means.

90

Morality, Personality, Mind

A continuing debate among biologists and others for centuries has been Nature vs. Nurture to explain why man acts and thinks and looks like he does. As it happens, we want it, and we actually can get it, both ways. Yes, man has genes inherited to make a certain brain, wired up in certain ways to make him very smart and social and all the rest. But, the environment, both the social and the physical (inner and outer) impinges on and profoundly affects the brain – starting as we saw, almost immediately after fertilization in fact. These continuously impinge upon his body, thoughts and mind to provide a means to endlessly modify both brain and thought and thus behavior. There are few human behaviors that have not been conditioned by innumerable environmental inputs and considerations, all the way back to *Australopithecus*.

As noted above, the environment can even directly affect the operation of genes and cause modification of their effects. This, in fact, is a major aspect of the development and life of every animal, most especially humans. The effect of culture on the developing and adult brain is, of course, profound – as witness the education enterprise. But it does not, and cannot wipe out the deep seated natural tendencies to respond in certain, generic ways to stimuli – like our emotions most notably. We need also to be mindful of Wilson's succinct phrase, "the biological mind is the essence and very meaning of the human condition."

So we turn to look now at the cultural aspects of the physical man that nature has wrought. For one thing, we are interested in how morality fits in with modern civilization. To presage a little, Robert Wright in "The Moral Animal" writes with grim good humor " these traits are good for society generally, yet they are not reliably employed for that purpose, in fact with brutal flexibility according to a manipulator's self- interest. In other words humans are a species splendid in their array of moral equipment, tragic in

91

their propensity to misuse and pathetic in their constitutional ignorance of the misuse."

His point is a most important one. Even though both selfish and good genes are in the human genome, that doesn't automatically translate into an overall good, or bad nature, or that we have to eliminate one or the other. We will in fact just have to live with both. The future objective is just to not allow too free a misuse of the "good" nature parts for bad uses – – which of course is what leaders throughout history have done and continue to do.

Also, as noted elsewhere, our genes are neither 'good' or 'bad'. Genes are totally neutral on that issue; they are passed on generation to generation by the inscrutable forces of natural selection. Good or bad is a moral type judgment and not within the realm of science. If there is one thing scientists – and writers and philosophers too – agree on it's that "Human Nature" is not always a reliable guide to derive an absolute universal moral code. It's not even that nature is amoral; it is that human nature has different moving parts and is fungible and multiple- faceted; at one time selfish, greedy, deceitful, murderous, or disgusting, the next loving, altruistic and moral. It is difficult to swallow anyone's simple assertion of any reliable truth given this ambiguity and changeability. Most scientists will assert however, that humans have developed, and fairly early on, their own way of exhibiting, at times, admittedly imperfect morality.

We have seen too, that not reason alone, nor ethics nor science, nor philosophy or art has come close to yielding a consensual moral code that has been successfully argued. In view of this absolute dilemma, the choice of a moral code upon which we could base a future foundation for improving the human condition becomes supremely important. It would seem only prudent and necessary that any of our newly chosen bases should therefore be based on the idea that the fewer and simpler foundational assertions the better. (I will later come back to this and discuss utilitarianism as maybe just such a code, as Robert Wright proposed twenty years ago.) Lewin several decades ago warned "the future of *H. sapiens* is now firmly in his own hands." That being the case, any consensual moral code we use should be the best we can do based in good measure by knowledge coupled with ample cortical reasoning and not dominated by our reptilian side.

Greene and Steven Pinker (2011) among many, many other scientists are making headway in studying the repertoire of actual behaviors which are innate, or most strongly programmed by our genes, both the individual, and group types. We have already mentioned some of these, like cooperation, selfishness, aggression, empathy, love, status seeking, dominance, compassion, etc. Many people have categorized basic human traits every which way, but let's look at just a few that make most peoples' list. Wilson, among others, has compiled a list of the evolutionary derived – innate – traits that have been studied. A few of these are: altruism, compassion, empathy, love, conscience, sense of justice, selfishness, fear, status seeking, deceit, aggression, dominance, introversion versus extroversion, tribal loyalty, optimism, hate, anger, disgust, greed, contempt, lust, approval seeking, avoidance of shame, guilt, embarrassment and a host of other emotions and feelings. Each of us could probably add many more to the list. These latent behaviors and personality traits are broadly inherited, remember. However, each one is not pre-programmed like a reflex; they can be and are moderated by the environment all the way from the beginning of life, childhood and throughout. Environment in fact, usually contributes nearly 50% to what our friends would say is our personality type. That, significantly, means that 50% is inherent as well, as I will explain a bit more later.

Out of these inborn tendencies of ours comes a form of morality. It is not a constant moral stance, but when his brain is "hitting on all cylinders" he reliably acts morally. Obviously, when operating emotionally and solely in the throes of high anger, fear, disgust, contempt and the like –

93

which is a great deal of the time – men can exhibit all the morality of a gorilla beating and eating their mate's young. It may not suit the intellectually rigid, but eventually we will have to come around to accepting this, at least pragmatically. (Along these lines it is interesting, that while the data are not bulletproof, they consistently show that in prison populations, atheists are less than 1% of the total, though they are 14% or more of the whole U.S. population.)

It must always be remembered also that these universal human characteristics are hereditarily determined through ordinary Darwinian evolution. Again, we don't mean that there is "a gene for that", like a gene for happiness, or gossip, or the like. There is no gay gene but there is in the brain, modules which, through the same intricate, incredibly complex brain development that occurs throughout our human behavioral history, have resulted in inherited protoprograms that eventually profoundly affect the final emotional and behavioral outcomes. Every trait can be modified to a great degree, but not erased. Can a person be cured of, say, selfishness? Absolutely not! But that person's public displays of selfishness can – at least sometimes – be modified. What is innate in all humans are the general modules and matrices, networks, and genetically organized structures that are acted upon and activated throughout the lifespan (again, cf. Damasio, 2010 and Sapolsky, 2017).

Relative to this, let me inject a little known – outside science – biological concept known as epigenetics. This well-established idea basically says that throughout the life of an individual there are constant environmental interactions impinging upon the basic cellular mechanisms. These then may modify to one degree or another, some of the complex cellular actions making up life. This includes constant interactions between the genes themselves and their surrounding milieu, which in turn causes the actions and expression of particular genes to be also altered. As one example, recently it was found that even the diet of a woman at the time of fertilization could produce changes in the gene

94

activity of the resulting early embryo. Meditation has also been shown to alter the activity of certain genes, as well as many other physiological and psychological activities. So there is little wonder then, that the life path of even a single person is endlessly variable and unpredictable. The basic lesson here is that environment and life's experiences have more or less direct, far-reaching influences even on a person's innermost physical, and psychical, functions. (The important recently started task of deciphering the "epigenome" will take a century or two more.)

It needs to be understood that things can be and are wired into our genes and brain – at least broad programs and algorithms. However, nothing in the Mind is hardwired. Is mind the only major thing in the universe that is infinite? Mind and morals are cultural concepts which seem to be naturally congenial and intuitively acceptable from the intrinsic workings of the brain of all people. It is not a big leap to say a man ought not murder his fellows; and this is built into a part of our brains.

These concepts lead directly to our legitimate use of 'ought', which is to say, a moral quality. We say we ought all have empathy for our fellow beings, or ought not kill our children. Humans, it seems, don't always have to decide to be empathic; by virtue of how their brains are put together, they just are (>95% of the population, at least are), discounting only psychopaths and a few others who are brain damaged in some way either before or after birth. But we need to be careful about our legitimate proscriptions of 'ought'. Society can legitimately legislate some applications of 'ought' in the realm of what people 'ought not DO'. However, generally not what they oughta, or 'ought not' think or believe. For example, the law *should* say that a Muslim – or any man – ought not kill his wife or daughter in some sort of faith-based killing. I should also add that Muslim men, or any man, ought not be allowed oppressive hegemony over women, or anyone, without consequences. Man has a ways to go!

95

Psychopaths and Sociopaths, A Digression

This may be a good place for a bit of a digression to talk about psychopathy/sociopathy. These individuals constitute somewhere between 2% and 4% of the population. (Babiak and Hare, "Snakes in Suits", state that the rate for CEO's is 5%!) Most psychologists regard empathy as a key feature in the human genetic psyche, and especially for morality and civilization. Hundreds of studies have developed measures for it that illustrate the ubiquity and usefulness, in social terms, of empathy. Turns out it is quite easy to test for psychopathy and/or sociopathy too, which are usually defined as extreme disregard for others' welfare in almost, but not all, situations; some even measure it by a nearly complete lack of ordinary empathy or compassion. Normal people (>96%) score quite high on the empathy scale. Psychopaths score below the 50% level, and even these can be differentiated into high, middle and low scorers, even though they all score very low.

Psychopaths have a characteristic detached psyche, which often manifests in actually an engaging personality, as they can fake emotions very readily. They are good at conning people. By these measures male prison populations are comprised of 20% or more psychopaths and they commit a large percentage, over 50% of all major crimes. I should point out though, psychopaths or sociopaths are not hardwired for violent or criminal behavior, but they are much more prone to many forms of antisocial behavior, for obvious reasons.

There is something inherently different in their brains from the beginning and this trait, unfortunately, has a high heritability score. In a recent study of physiological parameters within the brain, Decety et al 2013 studied a few dozen of these types of prisoners using fMRI. They found that 'normal' prisoners had clear, obvious and pronounced increased activity in the anterior insula, anterior cingulate cortex and right amygdala when they were shown graphic pictures of people being subjected to significant pain, like fingers being crunched in a car door, etc. Psychopaths registered little to no activity under the same conditions.

Another interesting finding (Pardini et al, 2014) is that their main fear and disgust center, the amygdala, is smaller than in normal people. This finding squares with their behavior in that being less able to experience fear might well make them more 'adventurous' and less susceptible to fearing potential harms or social disapprobation.

The sine qua non of psychopathy, starting always in childhood is an obvious uncaring approach, often bullying, dominance seeking, mendacity, cruelty to animals and unaffected by social disapproval. A key point though, is that psychopathy is clearly and unmistakably highly heritable. So to a lesser extent, is her brooding younger brother, sociopathy. Good twin studies make this perfectly clear. It is a lucky break that somewhere in our past this constellation of brain module characteristics did not become the norm, rather than the other way round. The recent book by Keysers, "The Empathic Brain" usefully reviews a lot of this area.

My point in making a special digression about psychopathy particularly is to highlight it as an especially egregious inherited personality trait that we have had handed down to us from our earliest forebears. This rare but powerful and unusual trait generates so many undue disruptions within a society that it merits special attention in the future. What, if anything, could or should be done with this phenomenon I leave to the future and more qualified journalists.

Of Sex And Tribes

It needs to be finally brought up too, the elephant in the room, i.e., the sex brain. Many of these behaviors are *the* most deeply seated. They arise mostly in the mid brain – the limbic system, hypothalamus and others – and many are of the instinctual bend. They most certainly had a very powerful

97

force in shaping early man's physical and mental features and have ever since powerfully affected social evolution. It was a universal force in forming tribes in the first place, as searching for a mate is the most powerful of all instincts in most animals. "Me Tarzan, you Jane". It has provided the main spark and energy for much of the whole human enterprise.

The cultural manifestations are so all encompassing that I am going to take the cowards way out and not treat this arena to any great degree. My only flimsy excuse is that everybody owns and operates one, so it is familiar. Somewhere along the line in the march of civilizations, sex became dirty, evil and sinful. There is much room for increases in societal happiness here, for example, by more extensive and enlightened education pathways. Millennials! Help the world straighten this out!

All In All, A Most Magnificent Ape!

As noted before, chief among the cultural and mental machinery of humans, ancient and modern is the need to form a group, and to cooperate within that group for the good of that group. Empathy seems to be a core trait that is the most pronounced of these proclivities. The power of the urge to form groups and then strongly favor people in that in-group has practically the status of an instinct. Within the group, there is a deep urge to seek approval and to avoid disapprobation. Gossip, for another example, it turns out is also a universal trait, and the topics of the gossip are also the same in all cultures. A sense of justice is another universal in all cultures.

Modern groups are the psychological equivalent to the tribes of ancient prehistory. Groups, or tribes gives status, social meaning, and make the environment less disorienting and threatening. Now our groups are friends, religious

98

groups, nations, ideological organizations, etc. The problem is we now belong, not just to one tribe, but a bunch of often fuzzy and interlocking, long range groups.

It is a tightrope that man walks, between his desire to fulfill his own universal wishes and his acknowledgment of social responsibility. The universality of virtually all of these common cultural traits should not be surprising since they arise from various brain modules which have some of the same essential, basic 'instructions' wired into each person. In any given instance, after this basic, deep seated program - often in the midbrain – is alerted, then all the higher processing, including finally the cortex, kick in to modify and/or obstruct a behaviorally orchestrated response.

A recent study beautifully demonstrated one of the physical underpinnings of some of these types of cognitive activity. Brain imaging showed that a particular activity in a specific part(s) of the unconscious brain was registering that a 'decision' had been made about an outside stimulus. A couple of seconds later, a correlated activity occurred in the cortical brain, indicating that a decision (the 'final' one) had also been made from the cortex. So, in this case, it almost looks like the unconscious brain first made the decision and it was just ratified by the higher centers. (It is assumed, though, that the higher centers could have overridden the unconscious, as has also been documented.)

There are countless other personality and behavioral features wired into the brain, a great many of them unconscious as Freud discovered a century ago. These common emotional responses we possess, some of which are good, some bad, sometimes causing us grief. Our understanding of them and ability to control them is also very weak. Emotional responses are generated mostly in the amygdala, hypothalamus and other limbic structures as we noted, but are radiated then throughout various sites – modules – in the cortex, and elsewhere too. How to *control* the natural emotional feelings, thoughts and urges in public action is probably the most pressing psychosocial research task and also the one that might give us more insights and help as we evolve (culturally). It is, however, perhaps one of the most dispiriting facets of modern society that well positioned groups of people, for profit or power or prestige etc., knowingly, and cynically, even gleefully use their cortex – higher brain – to play upon the midbrains of populations. They do this by use of language which is calculated to stir these natural limbic and amygdalar reactions of fear, anger, aversion, disgust and the like.

But we all know from long experience that emotional responses are to some extent internally controlled by people every day. Also, of course, emotional responses are not always bad either. (Good thing too, since they run us 90% of the time!) They are the main source of the most intense pleasures and may underlie moral intuition. While emotions don't always control our behavior, it is the most omnipresent and, in fact, underlies almost all of what we call motivation. Gets us out of bed and going in the morning to face the world. *Feeling* an experience is usually more intensely satisfying than simply knowing. Feelings are, in fact, most of life. Francis Fukuyama (2014) states that, in fact, all normative social behavior is grounded in the emotions and serves to promote social cooperation. He (along with P. Churchland, 2011) says that it is one of the great glues of society. Johnathan Haidt also argues that emotions ("intuition") provide the moral foundation of basic human life. This may be so, but our lives and society's are a daily play of the whipsawing of the two types of inherited brain actions. What ought we to do according to our two – multiple – brains?

It bears repeating---humans are *the* most highly social of all animals and the elements of personality that provided that impetus are still working, to some degree, in every human on earth. It is true that Man has made culture, but now culture is making Man. It is, in fact, a man- made and man- owned

100

world now. Putting our wisdom into political action ought to be a priority goal the rest of this century. As expanded on later, when a plethora of groups – tribes – layer their local morality over broader tribes, then the innate tendencies of Us vs. Them create titanic challenges. Reason, morals, imagination, art, justice: all these can and do evolve through culture and it makes sense to try to nudge the evolution using our better angels. Can we collectively transcend, even a little bit, our own human Nature? Fortunately, scientific data strongly suggests *yes*. In the next chapter I will briefly survey a few more recent studies of the inbred nature of humans and point out some possible dangers, as well as hope, ahead.

CHAPTER V. THE MIND AND THE FUTURE: WHO DO WE THINK WE ARE

The key question before us, again, is "what is Man and what is his nature". Who do you think you are?" What is the purpose of my life?" A character in Graham Greene's novel "The Power and the Glory" maybe had the best answer "it was intended for men the enormous privilege of life – this life!"

Questions like these might profitably be thought of in terms of what are some of the underpinnings as well as limits of man's amazing and increasing mental and psychic capabilities. It has been forwarded in some art and religious realms, that man has transcended reason, or rather invented a new realm based only partly, or even not at all, on reason or even on any known real form. In other words, there are other immaterial, or supernatural or other cosmic sources of thought or knowing (see e.g. de Chardin, 1955).

Thus, thought, feeling, mysticism, poetic flights, as well as true religious or mystic exploration are sometimes claimed to be new ways of knowing. Contemplation, prayer, feelings, or poems can contain equally true parts of 'reality' as in everyday life or the highest physics. They say these transcendent thoughts and feelings come from 'inside' somewhere, or 'outside' somewhere. Maybe from some mythical psyche, or the unconscious, spirits, or other deep, unknown sources. However, not from the rational, public, studied processes or the usual modes that people are familiar with. Many even say this is where wisdom and the essential true core of existence 'really' comes from and that gives us ultimate meaning and even knowledge. These are possibly true. But we also have discovered that a great many of these profound feelings can be faithfully reproduced at will by stimulation of known brain pathways and other known means too, as noted earlier.

102

Exploring the inner self, no doubt, can dredge out 'universal' truths that can provide much happiness and satisfaction, at least for that person. Meditators, for example, or mystics are often probably the most happy of men, as has indeed been indicated by many studies. "Mindful meditation" recently has become a rapidly growing practice in the West and has already been shown to have salutary effects on health and well-being both. Some businesses have reported solid reductions in worker accidents after institution of 'Mindfulness' sessions. In fact, future work on how conscious, studied exercises on controlling our minds may be our best future hope for the better world we long for. Teaching our mind new tricks has been the sine qua non of our whole human evolutionary path. As noted before, our minds consist to a large degree of telling stories to and of ourselves and how everything relates to 'me' and 'my' environment. This is how we make up and keep track of the world and "reality". Thus, teaching ourselves how to 'think' more beneficially should be duck soup for us. Sam Harris ("Wake Up") presents an excellent review of and guidepost through these kinds of mental exercises. All of these are, no doubt, of great value and hopefully in the future this will help us in our cultural evolution.

Writers (especially Sci-Fi), artists and transcendentalists of various kinds have imagined a future where teleportation or telepathy or the like is loosed from normal biology, leaving pure thought to and from the Universe. Shakespeare weighed in on this enigma: "Glendower: I can call spirits from the vasty deep. Hotspur: Why, so can I, or so can any man; But will they come when you do call them," (Henry IV, pt I). Literature and art in fact, is often more 'real' than reality and much more inspiring or beautiful to boot. The realm of feelings, imagination and even thoughts really brings us to some ineffable level, and science now falls silent here. Often these kinds of things fall into the category of 'spirituality', which concept appears to have come naturally to humans from the beginning. Indeed, some scientists –

103

especially cosmologists – are already theorizing and even pioneering in experiments around the notion that some sort of cosmological forces are indeed similar to or compatible with brain waves. Some have even hypothesized that we already could be part of some holographic simulation on a grand scale, or that we live in equally real parallel universes, or multiverses. Others imagine, and are actively working on how computers and minds could be synergized to make essentially a man-machine.

Information captured from the brain can, of course, be converted   into bits of information in computers; thus they can be transmitted indefinitely from there to anywhere throughout the Universe. The technical implementation of this is usually called Computer-Brain- Interface, or Brain -to- Brain - Interface or something similar. Several companies in this arena have started up and are well advanced. Dozens of patents have been already issued for equipment to accomplish this, both for medical and industrial purposes. Hundreds of scientific papers, and even now a few theses, are being published, illustrating dizzying progress.

I am reluctant to go very far into this new science literature; partly because I'm unqualified technically, and partly because it is too new to be really comprehensible. It will require decades to sort through and try to understand. But we can take a brief peep into the coming science that will blow minds for a long time. Time itself, and Reality itself are fuzzy notions, even for physicists (see, for example, Muller, 2016). Time, in fact, is not constant as everyone has always believed, but has been observed to stand still in some recent experiments. Gravity, for another example, is also a slippery notion as it has been recently suggested to possibly be an emergent property of 'entangled quantum matter' (whatever that may mean – for 99.99% of us anyway).

Photons have mass? I didn't even know they were Catholic.
You-Tube comment, John Doe   June, 2015

But, let's at least take a brief look at a couple of recent studies from various science fields to get bit of a feel for the future. Recently a group at Duke University collaborated with the Institute for Neuroscience at Natal, Brazil to demonstrate the first of a kind behavioral experiment in rats. Animals were put into identical experimental cage setups in the two continents and electrodes placed in multiple predetermined locations in each's brain. The rats were first given some basic training on some simple visual or pure somatosensory tasks, which when performed satisfactorily allowed them to pull a lever which delivered a reward. Electrical activity from the brain of the rat at Duke was sent, via the internet, directly in real time to the rat in Brazil. The result was that the particular behavioral stimulus seen by the rat in the U.S. instantly produced the identical response in the other rat (Pais-Viecia, et al , 2013).

Along the same lines, and even more mind-blowing, recently a German group (Folcher et al, Nature Communications 11 Nov, 2014) published this paper "Mind-controlled transgene expression by a wireless-powered optogenetic designer cell implant". They showed that by transmitting brain waves from a human, they could control the expression of genes in a live mouse. Amazing! A man's specific thought, instantly controlled a core aspect of life, which is the function of a gene – in a different species even!

The recent book on this new frontier by the physicist Michio Kaku, "The Future of the Mind" also may give us the shudders, but is nonetheless interesting reading for a glimpse into some of the fantastic future. tyProjects like genetically engineering quasi- immortality – e.g. significant extension of life; or core body part replacements, or thinking robots, which are already nearly feasible (see for example, Montecucceo, 2006; Kotler, 2015). In fact, a robot has recently been reported to have passed an initial test for self-awareness. Cyberimmortality is being spoken of. (However, we should not forget that many organisms, like our kin the jellyfish, for example, are already immortal.)

We already "know" from science fiction about how such scientific advances could proceed – – for instance, Scotty beams up Captain Kirk, or Microbots prey on humans in Michael Crichton's "Prey" and so on. The marriage of neuroscience, nanotechnology and robotics will be a very powerful force which could lead to, for example, putting something like the Connectome on a chip – which would be tantamount to having a disembodied person on a chip. And what can be made of the possibility (and as Kaku argues, this is feasible and still obey the laws of physics) that post-biologic individuals could be liberated from any solid physical form, i.e., all pure consciousness. "Transhumans", cyborgs, and the like. Exciting, but scary stuff. The situation is tantamount to admitting that the mind is virtually illimitable, such that it could, in fact, lead to some kind of an "ultra-homininization", or a new level of hominins – the Mind only, "Mind-Oids" ? It certainly does look like some kind of positive feedback loop has occurred. But any positive feedback loop, by definition, means that something increases forever – or until some negative feedback intervenes. Eventually could man not figure out how to make robots fully autonomous and capable of making smart, fully informed decisions, i.e. 'aware'?

This is what people are afraid could happen. Already over a decade ago, McKibben ("Enough", 2003) particularly warned of the dangers of the coming scientific advances in biology, robotics, artificial intelligence, materials science and nanotechnology. Mckibben reviews a good deal of the scientific fields that we have briefly touched on, specifically with respect to rapid advances in these neuroscience-computer interface and genetic engineering arenas. His main conclusions back then, urging vigilance, seem valid still today. However, some of the current fear stories of bots turning the whole world into "gray goo", or Cyber-Man soon running roughshod over our poor remaining biologic humans and the like are still a good bit hype.

106

As is the near prospect of immortality and super-intelligent human clones selected for multiple superhero traits. Nevertheless, he warns, correctly, against too readily barging ahead with "mangling of humanity and the destruction of the meaning of our lives. In other words, what would these enhancements be for?" [I'm not at all sure what it even means to ask "what is the Mind *for*?", except from the evolutionary perspective. Again, Dennett (2017) and also Damasio (2010) go a long way toward explaining this process.] Also, a recent book by Diane Ackerman ("The Human Age") gives an interesting, intelligent and comprehensive tour of the exciting and scary progress in a large number of such arenas.

Game-changing revolutions are going on in almost all the sciences, but many of these changes are highly positive, i.e. they will provide immense improvements in the lives of people. Unlimited free energy, for example, unlocked from the quantum world is quite likely on the near horizon. Already, experiments are going on around the world which, if in a few years turn out successfully will be a great boon for society.

Actually, this whole arena of high science, especially cosmology, quantum mechanics, and digital technology boggles the mind of the greatest scientists too and are really beyond any present human understanding. Dr. David Reilly in his talk on 'Quantum Nanoscience' at the 2013 ISS Conference in Sydney stated the view of most physicists: "If you think about these things too long it will make your head hurt." Virtually all physicists would wholeheartedly agree with him. Richard Muller in his new book (Now: The Physics of Time) gives a thorough, erudite and eloquent overview of high physics and its relation to society, the mind, and even the 'Soul'. He says this stuff "can sometimes drive you crazy".

However, we should remain mindful of the fact that our brain is presently our only known transduction mechanism of reality. It is, remember, a real, evolved physical mass.

107

Miller (1981) says that in fact "The crowning intellectual accomplishment of the brain is the real world." Of course, our brains, no matter how powerful our tools, can never possibly comprehend any more than a limited slice of reality and it will be forever impossible for *complete* knowledge. It would be silly for us now to imagine otherwise. The common question of what is actually "real" is, in fact, unknowable, either scientifically or philosophically. We cannot ever know the secrets of our 'real' existence, or the limits of the Universe and many of the other puzzles, arguments and questions posed by philosophers, scientists and mystics over the ages. Our ape-brain will have to suffice – – for now.

While I'm at it, let me briefly summarize my rather staid, run-of-the-mill stance on these mysteries of life (or the meaning of life, or the mystic vs. the scientific etc.). First, God, or gods, are definitely out. There does not seem to be room in the whole immense Universe for a God-buddy who can watch over everybody 24/7 and have his hand in fine-tuning everything in the cosmos all the time. Why would an all-powerful god be hypermoralistic, judgmental and obsess over and keep tabs on every private deed, thought or eating or sexual habit; one would think there are lots of other more weighty issues somewhere on this planet or around the Universe. If the Creator's person or spirits are somewhere in the cosmos floating around, sometimes visible or made known to us in unusual ways, I would be anxious to know more about them. The Creation myths and other spiritual dogmas which all religions are founded upon, seem silly on their face and insistence on faith in them in the public arena seems cruel.

I accept that people sometimes seem to have favorable and profound experiences like these but I do not accept that they have in fact conveyed essential truths of immense import for all mankind. I also agree that Man has something we call his 'spiritual' side (whatever that may mean.) Certainly there are things which are totally, now and forever, unexplainable,

108

e.g. thoughts, feelings, and unguided and unbidden bouts with an unknown conscious basis, for example. Especially *feelings,* like the ineffable wonder of, beauty, awe, joy of life, or the color blue, etc. But I am more than skeptical about assertions that these are purely 'spiritual' from somewhere and have no source or mutability.

On the other hand, Science does not, and will not, know everything a human could conceivably want to know. There are forces and 'things' on this planet and in the Universe we cannot perceive or fathom, maybe like dark matter and energy, for some simple examples. We do know that we are not alone, though, as there does exist, in fact, 'things'. There is definitely an absence of Nothingness. We accept that there are things like billions of galaxies, black holes, undiscovered species, electromagnetic energy, gravitons, etc. The Universe has a lot of weird stuff that we have been successful in actually observing, and also a lot of things that we either think might be there or could concede might be there.

Thus, we are still faced with the final frontier, the big mystery; why is there Something? We here, the pinnacle on this planet are made of this 'stuff' whatever it all might add up to be, so indeed we already are ' One with the Universe'. Indeed, I take it as an established fact that not only is all Life 'one' but it is also 'one' with the Universe, from whence it came.

I don't know what the ultimate meaning, or purpose of life is. Nobody does. Nobody ever will. The primeval cry, the basic 'terror of existence' is probably mainly the fear of death and there is no totally satisfactory cure. All we ever can know is "what is the value or purpose of my life", today and for the foreseeable.

That is why I think it so essential that we try to understand how we can fit in with our own environment and especially our own precious human species. If there is no satisfaction or pleasure or happiness to be derived from our present life, our surroundings, our fellow man and our own internally derived purposes and goals, then we really are in a bind. At the least, we can conclude with Descartes that we are certainly along for the ride, so why not try to enjoy it?

A few final comments may be in order regarding our scary technological future. Things are clearly still speculative and possibly not a near danger. But the current status of science is surely indicative of what is to come. Most likely it will eventually present very real, very great dangers indeed. It would seem prudent to conclude that no enterprise which has no finite ending should be blindly allowed to proceed infinitely. In any case, this future is still off a little ways so we must still keep on keeping on but keep a close eye. The dangers will constitute one of the issues that we will, in the next chapter, set out as one of the tasks for our new governing machinery to also solve. We will have to control our own increasingly dangerous technologies if we are to remain human. I want to make clear – – science is not The Truth, nor does it give out *Truths*. It too, is based on a type of faith. We have faith (mostly 'trust', actually) that what our eyes, or instruments tell us really is out there, and is real – some of it anyway, and that often darkly. And we act on that information accordingly. Of course, if that 'faith' turns out not to be in line with our other barometers of reality, then the faith must be rejected. We then can make hypotheses and experiments and see to what degree the phenomenon agrees with us, and with replications and with other people. Hypotheses, however, are just *ideas*, i.e., immaterial as far as we know. Intuition is a big part of science too, just as in any other endeavour. Science obviously uses ideas, as well as metaphors, subjective reality and all the rest, and just as in any other human activity some of this leads to beliefs.

Beliefs by definition are believed to be true, whether they are or are not. In fact, it is impossible to believe, logically that all beliefs can possibly be true. To a true scientist – or anybody for that matter- a belief should be tested in various ways to ascertain how likely it is to be true, or under what conditions. Beliefs, whether based on data or not are

110

fungible, and as experience shows, very often not justified. Since not all beliefs about the world are settled, or settleable by science, then experiences, reasonableness and common sense must also be exercised. Beliefs, as the philosopher Sam Harris in "The End of Faith" states, "are tightly (usually) coupled to the apparent structure of the world. Is a person really free to believe a proposition for which he has no evidence? No. Evidence is the only thing that suggests that a given belief is really about the world in the first place. We have names for people who have many beliefs for which there is no rational justification. When their beliefs are extremely common we call them 'religious; otherwise, as Harris (2004) says, they are likely to be called 'mad', 'psychotic', or 'delusional'. What can we say about the beliefs of skeptics that, for example, deny the age of the earth or evolution, etc. As Adam Gopnik (The New Yorker, 2014) said "The plausible opposite of 'permanent scientific explanation is singular poetic description ."

Elsewhere I have emphasized that scientists are people and they fall easy prey to all the foibles, nastiness and cruelty of everyman. But since many scientists are also 'elites' and knowledge -(data)- gatekeepers with actually a lot of collective power, their tribes could, like political tribes or religious tribes become a dangerous lot. But, notwithstanding, none of this excuses Man from using the methods and the reasoning of science and the careful descriptions of our shared sensual reality in his social affairs. Nor does it negate what we call our common reality. Our highly evolved reason is a thing not to be lightly regarded or trifled with – as it presently is in many quarters. Reason in service of government is a high priority in the future, as I will delve into a little in the next chapter.

I will leave it to more qualified minds to forage in this rarefied scientific-philosophic realm and make it more accessible. It may take another century or more to know if there is actually anything to this whole sci-fi type scenario, but certainly it is prudent to not close the book on it. In either case, it will not lessen the need for the transcendent, the infinite, the soulful striving beyond ourselves, from the arts especially that is so critical to a satisfactory humankind and a better future for him.

CHAPTER VI.   WHAT NOW

After this whirlwind review of the history of the man and Mind, it might be decent to make some effort to help as we humans grapple with the future. Maybe to claim even a minuscule amount of credibility as a guide through the mists over the next century I should summarize some of what I am trying to say in this book, and try to develop some kind of guideposts for real people and realpolitik.

1. Man is by inheritance (nature), a tribal animal and is primed to act selfishly and competitively by large segments of genetically conditioned parts of his brain.

2.   Man is also primed by evolution to override his "selfish" genes and feel widespread empathy, love, cooperative, altruistic and similar social urges in other networks of the brain.

3.   Man is primed to act cooperatively, altruistically and loyally   firstly to his own in-group and act competitively and/or hostilely to other groups. (This does not, however, preclude evolution in thinking to modify these behaviors.)

3.Men, like wolves or ants are not inherently moral or immoral, good or evil, religious or secular; these are invented terms used to characterize his social, mental and cultural behaviors.

5.   Man has created psychological and mental constructs that he uses to guide his behavior, irrespective of inborn tendencies.

113

6. Man's Mind and thoughts are potentially infinite and are his own. When transmitted by language, man's thoughts are potentially permanently magnified everywhere and becomes part of Everyman's changing environment.

7. By virtue of man's magnificent and comprehensively successful evolution to become king of the world, his social governance assumes critical importance to keep the enterprise going.

I have, quite superficially and speedily, introduced material relative to each of these 7 points, except for the last one. Actually, all of human civilization relates to #7 in a way. In essence, the point is: how can we keep tribes from oppressing or killing the others? We traced the human journey through 'real' history the past 10,000 years or so and one thing we can see is the jagged and painful path we took. And yet it seems to have a clear arrow upward trending, ever so slowly, painfully and erratically. It has been a series of veritable laboratory experiments constantly bubbling up, each with clear marks of the struggles within the man and between Me and Them. Now, as in the past though, the struggle is often posed simply as good vs. evil, which generally is not helpful.

The "I", the individual, particularly stands out – as history students know, it seems they must learn one king's name after another! The apparent psychological necessity to differentiate status, i.e., lust for power, has more often than not trumped some of the other human urges of cooperation, love or empathy. One can hardly know what to call the cruelty and implacability of most of the early – and current – empire builders for power and space, or the submerged agony of the masses caught therein.

Again, it is interesting to speculate on exactly the number of wars, big and little, that humans have fought over the millenia. The Durants ciphered, somehow, that "in the last 3,421 years of recorded history only 268 have seen no war". Pinker in his latest book "The Better Angels of our Nature" provides details about a large number of those and surveys the entire history of violence. He quotes sources that say that there were at least 2300 wars in the past 2000 years. In the book, he also argues persuasively the point, an important one, that human violence has been declining for centuries. He optimistically cites some cultural adaptations that probably account for it.

With each turn of the wheel of time then, one might, as I also do, optimistically see signs of evolving culture and maybe also a relatively new enshrined mental artifice – – Hope. The Greeks helped, so did the Romans. So did Charlemagne, and so did Michelangelo and da Vinci and Galileo. So did the 16th century and the 17th – Columbus, Gutenberg, Botticelli and Newton and Shakespeare. Up through the Renaissance and then shortly, the Industrial Revolution.

Ah, the Industrial Revolution. It started let's say, in the mid 1700's with the invention of the steam engine; then shortly mechanical weaving machines and in the 1800's the railroad, telegraph, cameras and steel and big, huge machinery. After 1900 the aeroplane, radio, fast autos, stop lights, WWI, WWII. And then it hit the fan! Modern! And what does modern mean? (At least that is different from what modern meant to the well turned out Roman?) Not even modern, Postmodern now.

It is ironic that some of the postmodern Man has got so tied up in knots and into dystopia of all sorts by the blinding speed of change. But we are, remember, dealing in the end with a still active caveman inside, with whom existence and simple things are most of the game. A universal anxiety arises in each young person – their whole generation in fact – as they grow up and contemplate how they can make their way in the world. And it is bewildering, with few sure inner – or often outer – guideposts. Internally, we are actually

pretty well equipped to work in society, fall in love and, if
our environment be not oppressive, live quite successfully.
This is what we were evolved to do, with all the emotional
and intellectual equipment to carry it off. This is the core
game of humanity. But if the full force of bewildering,
swarming and hostile society falls piteously on a young man
and there is little help or hope available, life can get
hopelessly confused and hard; joys or options become few
and far between. (Human social life was always a tough job.)

   Speaking of generations, the whole notion of looking at the
life of nations through the prism of generations and their
unique life experiences is an important one and is well
explicated by William Strauss and Neil Howe in their
intriguing book "Generations". There have been typically
six generations alive at any one time:
Very young, Childhood, Young adult, Midlife, Old (plus a
few Very old). Altogether, this mix leads to the society as it
exists at any one time and how it will continue in the near
term at least. For thousands and thousands of years the
young generation has marched on stage like gigantic ant
columns into each little window of time – actual life – and
the elders tumbled off the stage into the mighty river of the
past.

   This, the greatest drama of all is played out by each
generation. Young men and women leave the nest and set out
to make a nest of their own. They, most of them, naturally
fall in love and more or less unconsciously but naturally
form their own sacred units we call families or communities.
Around this core they carve out their way in the wider world
somehow or other as events and opportunities present
themselves. As time then passes, they accumulate memories
and the offerings of their society as well as they can manage
according to their gifts and circumstances; thus their lives
pass. Few realize or appreciate at the time the greatness of
this drama of their lives. Only later do some see the
powerful, yet fragile homely satisfactions that occurred
throughout and – hopefully – gave, and continues to give

meaning to their lives. This, in fact is what they lived *for*.

Of course, for many this progression is too much filled with drudge, drear and despair. Whether from their environment or demons of all sorts, little satisfaction or well -being characterized their lifetimes. The remedy for this is still being sought by societies – some of them anyway – , and then only sometimes. It is, however, becoming increasingly clear to science that the brightest path to optimal, or at least more, happiness or well-being lies in enriching the environment (this is equally applicable to humans and animals).

The famous "Rat Park" experiment over 30 years ago by Bruce Alexander and colleagues showed clearly an overwhelming role of 'environment' in animal life. Alexander noted first that rats housed in standard rat cages – solitary confinement – showed many subtle, or not so subtle, effects of stress. When these rats were given free access to morphine or other drugs, most appeared to become hooked (analogous to human addiction) and strung out, usually lethally. Alexander then made a new kind of environment and repeated the heroin experiments. They placed a small number of normal rats in a large, very nice upscale suburban type environment (the rat park). It included rats of both sexes, toys, lots of room to explore and hide in and altogether behave like free rats. These rats were also offered, ad lib, the same drugs. But in this more congenial environment most consumed very little drug and they remained healthy. There were few drug related pathologies noted and no addiction. This indicates that something in this rich environment prevents drug addiction. Next, they took some individually caged rats which had previously become hooked and placed them into the rat park. Most of these after a brief time quit drinking very much of the drug and they eventually became healthy again. Other scientists have shown this same thing repeatedly. Johann Hari's new book ("Chasing the Scream:The First and Last Days of the War

On Drugs", 2015) provides a nice reprise of and context for these stunning experiments and concept.

A considerable literature is building for humans, too, which points to the same kind of conclusions (see for example, J. Newton, 2007, "Well Being and the Natural Environment: A Brief Review of the Evidence"). It is, in fact, the main purpose of this book to encourage humanity, as its greatest unfinished challenge, to develop ways to impart a strong upward trend to the well-being of future families through improvements in their social environments (using their magnificent, evolving, hard-won Mind, of course.).

Traditionally, the child and youth generation, as well as the old – and very old – generations have played quite limited roles in the active economic and political life, while the rising adult and adult generations play the major roles. This natural situation has created its own unique generational dynamics and tensions thereby. Actually, people in midlife are, legitimately, pretty much totally immersed in and fully occupied by their busy lives. They are fully engaged in making a living and enjoying life in the fast lane, or at least aspiring and yearning for the material and psychic rewards normally attaching to vigorous life-loving adults, not to mention raising, educating and nurturing children. This means that the governmental and other organizations needed to do all this has to be in place and running more or less on autopilot so that these activities can proceed apace, even as most of the energies of two of the four generations are 'distracted' by living life. This makes harder the task we will set us to do below.

One particular generational problem is that smoothing the generational transition – for example, our current one with the Millennial generation – is not even attempted in any discernible rational fashion. That is, the Millennial generation sees our present society, which is built upon this century's worldview of money, more money, exploitation and the like, as being trapped in terminal fail mode. They

118

Sense that their life aspirations will likely never be satisfactorily met by this kind of society. Worse, there is no visible attempt to accommodate their point of view so that their world would actually be better. Some people, blithely and wrongly (see Grabow, 2009, "Voting With Our Pants Down") call the Millennials themselves civically illiterate, as for one thing, they are accused of voting at low levels. Personally I believe that within 25 years we will be thanking their whole generation for starting to more satisfactorily sort some of this basic kind of stuff out politically.

So where could we go and what we could do now and for the future to aid our young man's entrance into and then happy existence in the world? One thing we could do is try to keep in mind a naturalistic understanding of our evolved human nature and its morality lessons.

There are no absolute precepts or sure judgments here, but there is lots of room for warnings and modifications in social processes. We certainly do need more reliable, consensual information about psychology and sociology to build on. This is not to discourage the arts and letters and all the rest of human thought and endeavor; as indicated earlier, this is where humans *live* and will increasingly in the future. Science alone, no matter how sophisticated it gets, is not nearly enough for a full, truly human life.

Politics

Therefore I urge, maybe a bit churlishly, that what the world needs now is a good 5 -cent political scientist. We need to acknowledge that we cannot within any reasonable time frame change our inborn brains, with all its spring-loaded behavioral repertoire. We also know though, that it is entirely feasible to change, soften, or damp down some of the more or less automatic responses acquired through cultural evolution. Thus, we could make progress over reasonable human generational time frames. But the best real hope for near and intermediate term progress on the macro level seems to me to lie in simple, rational governmental tools. Many of these are already extant in the world and are really no secret. But we will need to develop improved structures and institutions and paradigms and methods to keep some of our tribes from killing each other – or the whole species.

No longer are governments mere tribes. More extensive tools will be needed to have governments actually work in this large, relatively new realm. Governments are a recent invention as we saw with the rise of civilizations. We will need to solve the energy crisis, environmental crisis, poverty crisis, religious crisis, overpopulation crisis, economic crises, and on and on. But these are all related, and difficult though they be, they are all man-made and therefore all are man solvable. Politics is the art of making choices and is the main mechanism by which worthy societies integrate all the ideas and choices and keep society together. Actually, we should say that political science is part science, and part art.

There must be useful ways we could approach this. As we saw earlier, there have been throughout human history very few avowed political scientists, and even these have had minimal impact on their compatriots or us. Machiavelli might be pleased to know that he may have invented the genre, but I doubt if many would be willing to follow his lead too far. The names we remember are few; Locke, Paine, Hume, Bentham, Mill, Adam Smith and a few others. They have had considerable influence, especially on our American founders, but now there are no clear concrete theories or guidelines that humans will rally toward in future. We might even say that political science has not progressed far and may be "the dismal science". It is not as though people have not thought of a lot of ways to make better governments and social institutions. Any day of the week brings several books, multiple essays, blogs, articles, studies and speeches which contain either a little or a lot of, often very good, specific ideas and means. However, except for Thomas Paine and the

120

founders long ago, little meaningful effort has been initiated to actually build a better mousetrap. The greatest single contribution so far to governance, political science and even consensual contracts themselves comes, I believe, from Paine who clearly stated the principle that wound up in the Declaration of Independence; the only legitimate source of governments comes from and by the people themselves.

I claim no great new specific ideas or programs and cannot myself provide a whole lot of help here except only in really quite simpleminded ways to point to a few possible starting points. Maybe we could have a long term program like the Manhattan Project. One, however, which is decentralized, emanating from a multitude of sources and envisioned as a synthetic 'government engineering' project designed for the long haul. (I might also, humbly, suggest that early on, the new 'doers' recruit along the way a good ten cent economist to 'splain what money is, how it is made, and how much of it there is, or could be – – and what it is for.)

In order to invent a better human political order that will facilitate, or even allow, solid progress toward the goals that are dreamed of, we will need, though, some new concepts and tools, especially tools. It would seem foolish for the modern political scientist – the next Mill – to start designing guidebooks for the government of the future using just the technology and social structures and mores of the present or past. Continuing the obsession of governments since the world became hot, flat and crowded after WWII, with the mantra of industrial growth, jobs, markets, unregulated individual or tribal economic competition, etc. will not do the trick anymore. It would be no more successful than the prior millennia's fixation on oligarchy, monarchy, religious, or military juntas, or personal or empiric aggrandizement under any or all guises.

The economist Paul Mason has recently offered very cogent arguments that the end of capitalism has begun. He says that

121

we are already inching into the "post-capitalism" era, or rather into a 'non-capitalist' society and offers some sensible suggestions for future governance. Very recently, Kate Raworth ("Donut Economics: Seven Ways to Think Like a 21st Century Economist", 2017) also offered a whole new way of looking at money, economics, and society. She especially skewers the age old dogma of all-in pursuit of "economic growth" as the holy grail.

What to do with eight billion people is the number one reason for beginning some kind of new economics. The civilized world, especially the New World since the Renaissance was lucky. The population of the world in 1700 was about 650 million, and only 250,000 in America. But of course, the usable land area was the same as it is today, about 2.3 billion acres in the U.S. All of the new and expanding human capital and most of the land riches – which was everything in terms of wealth creation – was still empty and available to exploitation. With the tools and supposedly illimitable land, it was possible and deemed desirable to grow and spread and prosper at max speed with a free and heedless hand. This extraordinary goldilocks convergence of space, time, opportunity and chance is over, and the tools and philosophy will not come again. This attitude and approach which was made possible by empty space and was deemed inevitable is now judged to have been stupendously successful.

Now we have come to the end of it! Now what do we do? What are people to do, when labor in the old sense will soon be all but obsolete due to technology, and there is no room for hunting and gathering or even recreating in old, familiar haunts. If menial, or even most industrial jobs per se are going to be virtually extinct (think robots, AI, 3D printers, etc.) what are people to do to be truly human? (Does anyone really believe that menial 8 to 5 jobs will be the norm and expectation in the next century?) Education, tourism, recreation, entrepreneurism, environmental restoration and entertainment may grow and morph generation after

122

generation and provide meaningful occupation. Politics and public service may also be one of the great employers – lots of police, analyzers, watchdogs, bureaucrats, Army, and others will still be needed. And education could employ very advantageously a great number. But what are people to do to be safe, happy and human is a continuing question.

John Kenneth Galbraith in "The Affluent Society", an old and dated but still worthwhile book pointed out a number of salient features of historical societies and nations. The major point was that heretofore every nation was poor and most people living in poverty. Most people were living from hand to mouth, at or barely above subsistence levels. This was the norm for almost everybody, with the exception of a stable handful of rich landowners, princes, dukes and the like. And this was the accepted, natural way of the world. People were always as he says in a state of "peril and hopelessness." His book is an expose of how these old systems of thought – conventional wisdom – are not only around yet, but shape still so much of our affluent society.

Foreshadowing a related topic to come up a bit later, here's a little example of a tribally derived moral stance in our current politics. It illustrates the dissonance between deeply imbedded tribally derived moral stances and more rational and reality based behaviors in our post-modern "Human Age". I refer to the widespread conservative position that nations must first, above all, see that the individual's self-interest seeking and wealth accumulation should be the primary governmental and societal concern. This process, they say, must remain virtually completely unimpeded, to let the prosperity of the few "trickle down" to the majority.

This notion, however, seems more like typical tribal catechism, impervious to data or analysis, than defensible, sound organizational policy. This lodestone theory has also been pretty well objectively demystified and found to be deficient generation after generation. Thus, it would seem to be clearly unsustainable and an unwise and immoral governmental stance for the public at large in the future. This

123

is just one example out of dozens one can find in all tribes, liberal and conservative; all have their 'blind spots'. (For some amusing results of trickle down, see B. Ehrenreich's "Nickel and Dimed".)

This may be a good time to dust off another old canard: that humans are by nature just savages with a thin veneer of civility or morality: or "humans are animals wrapped in a desperately thin cloak of culture and civilization". By this ideology, humans make war because at heart we are aggressive animals; we arose from bloodthirsty ancestors so that is what we do. Rape, violence, poverty, cruelty, selfishness, all the many blots on human society are rooted in human nature and therefore nothing really can be done about them, no matter how loudly we might point to our civilized veneer.

All of these are − partly − true. As we have seen, it is true that some of our brain is devoted to greed, selfishness, dominance, aggression, rage and fear. Darwin himself in his book "Descent of Man" cogently made similar conclusions about our inherent dual nature. He also, however, emphatically noted the panoply of good genes, like love, cooperation, compassion and the like, as a strong part of human nature (although, of course he knew nothing of genes). The biologist Gould ridicules the savage beast inevitability argument,

"How convenient to forget that, as the only ethicizing animal in the world, such decisions are ours to make; they are NOT written indelibly in our genes."

Empathy seems to come the closest to being the most natural moral behavior which we could use in our future quest for better governance. It is seen especially strongly in chimps and bonobos but occurs too, in most mammals. While fickle and limited, empathy is such a ubiquitous trait that if the reasoning brain is brought into it as well, could be

124

near an ideal vehicle of impartial morality. Below we will look at a couple of other examples of how ethics and morality probably arose from nature itself and, therefore, why, civilization is not just veneer and doomed to failure.

Moral and social impulses arose in social animals millions of years ago so there should be no outraged surprise that humans should have these same traits embedded in the brain. For example, Bekoff, a zoologist who has long studied wolves, writes regarding a wolf pack: "During social play, while individuals are having fun in a relatively safe environment, they learn ground rules that are acceptable to others. There is a premium on playing fairly and trusting others to do so as well. There are codes of social conduct that regulate what is permissible and what is not permissible, and the existence of these codes might have something to say about the evolution of morality." Amen!

People who see our nature in Hobbesian terms and only 'red in tooth and claw' and our desires as all bad, while partly true, miss the whole point of this book and most of the last couple decades of some good field and experimental science. Hopefully, we are approaching a time when Thomas Jefferson's vision becomes true, when he wrote shortly before his death "The general spread of the light of science has already laid open to every view the palpable truth, that the mass of mankind has not been born with saddles on their backs, nor a favored few booted and spurred ready to ride them legitimately, by the grace of God."

Lessons From Our Closest Cousins

It seems useful for a brief digression here to look at our closest living relatives, the chimp and bonobo, as it could aid in eventually designing new societal relationships for humans. Both of these animals are smart as a whip, both are very social, both have very similar habits and habitats and they illustrate their – and our – two natures very well. Their social natures, however, are quite different from each other.

125

Goodall and others have extensively studied chimpanzees and Hare, Wrangham and their colleagues along with Kano and Frans de Waal especially, have recently studied the pygmy chimps, bonobos. These latter may be genetically closer to humans than chimps are to humans, though we do not know this yet for sure. Both the chimp and bonobo genomes have been recently sequenced, so very interesting information will soon be coming out. Bonobos and chimpanzees became separated from each other as distinct species only 1-2 million years ago, and both share DNA with humans up to 98.8%. (Compared to 98.4% for gorillas. Interestingly, bonobos differ from gorillas more than they differ from humans. (For more comparison, individual human DNA differs around the world by only about 0.1%.)

De Waal and others have described the bonobo social mind as characterized in the main by cooperation, empathy and sharing, and not aggressive or violent. Unlike chimpanzee society, which is often violent, hierarchical, murderous and domineering. Chimpanzee social groups are usually dominated by males, although females, at least the high ranking ones, often are aggressive and controlling too. They have a strong hierarchical organization in any case. Intra-tribal homicide or violent attacks are not real common, but inter-tribal warfare is quite constant. Killing of males especially, but also young and sometimes females is always a possibility. Inter-tribal cooperation is not often observed. But at the same time, chimpanzees in the normal course of life are calm, loving, altruistic, cooperative and consoling, often sharing grooming and sometimes food. They often exhibit high levels of empathy, compassion, sympathy and deep loving bonds between a variety of individuals – not usually however, with individuals outside their tribe!

Frans de Waal describes one example of this moral urge where a colleague was walking through the grounds of chimpanzees on a dark rainy day and noticed a pair of chimps huddled together cold and miserable in the rain. The door to their enclosure had accidentally gotten shut, so he

126

went over and opened the door. Before they hurried inside, both of them stopped and gave the man a hug of gratitude!

In other words they are quite like humans, with a generally cooperative, empathic, even loving disposition to a considerable degree. However, their other persona is hierarchical, domineering, aggressive and violent, which is exhibited in quite a few other occasions. As noted earlier, it has been estimated that up to 30% of male chimpanzee mortality is by murder. It should not be understood though, that chimpanzees are barbarians at heart, all things considered. Much of the time their other social proclivities, morals, and culture are displayed, sometimes gloriously. (Killing is done mostly by raiders, – war parties, not armies – and sometimes by an out of control male. Sound familiar?)

Bonobos, on the other hand, have a female leading the group. It is the sons who stay with the group for life while it is the females that leave and join other tribes. There is a minimum of fighting and virtually no warfare, either with -in or with-out tribes. The bonobo society is characterized by an easy going, non-hierarchical lifestyle with play, sex, sharing and cooperation the overwhelmingly frequent activities. There are conflicts of course, but they usually end with play, sex or food sharing rather than bloody fighting. Bonobos are like humans in many of these aspects, lacking only strong aggressive, domineering warlike urges. Society could learn much from these remarkable animals.

The neuroscience studies are just beginning on these groups, but already interesting observations are available. For example, Rilling et. al, 2011, at Yerkes Primate Center performed anatomical and neurophysiology experiments on both chimpanzees and bonobos. They found clear differences between them in, for instance the anterior insula as well as other sites. These were correlated with different, specific behaviors. Many of these kinds of studies, including detailed genetic studies, have been published by workers all over the world. The main takeaway from the bonobo- chimpanzee studies and comparisons is that their brains operate very similarly to each other and to us. None of their innate behaviors are substantially different from ours and the internal activities within the brain will likely turn out to be mediated by the same structures in largely the same ways too. Bonobos tend to react one way, chimpanzees a slightly different way, and humans yet a somewhat different way, but all clearly closely related.

Coming back to our own human brain, we can predict that, because of the way it evolved and has so many disparate, opposing and contradictory spring loaded emotions and behaviors, it will be impossible in the foreseeable future to design any one social scheme for living to ensure happiness for everyone. Everyman has in him the need to be king, to have more than most of the others, and at the same time to be loved by everybody and to do good, etc. We all also even seem to have the need sometimes to just go rogue, and create momentary unhappiness out of frustration or spite – or something. Simple living, with its inevitable unbearable tensions, despair, conflicts, pain, disappointments and the like can sometimes get the best of anyone and create temporary havoc. In addition, quite aside from social and political ordinary life, humanity lives partly in a mythic, spirit-haunted world, for which religion and superstition have historically been the main carriers. But again I emphasize, we have made this history ourselves; it was not thrust upon us from some supernatural being or force.

To get around the problem of our inherited mixed up brain it will be well to consult our better angels and our sounder science and wisdom. As Harris, Pinker, Kahneman, Dennert, Damasio and almost all current biologically oriented scientists and philosophers agree, there is plenty of room for mental evolution to take place rather rapidly. There ought to be ways to organize local and national groups and institutions in ways that minimize viciously destructive conflicts and leaving room for people to freely strive for their individual goals and professions, etc.

Some Modest Proposals

It is certain that there are some practical ways to at least make some kind of concerted action for us to make progress. Let us turn first to an old source, to a resurrection of John Stuart Mill. He is most famous for his espousal of his teacher's (Jeremy Bentham) utilitarianism. This is best known by the epigram 'greatest happiness for the greatest number'. Basically Mill called for a new paradigm for designing a governing system, i.e, a new political science. His first suggestion was, like Machiavelli, Hume, Locke and others, first use logic and reason and science to design governing systems without getting mired down with ethics and religion with all their baggage. The question then is, what are some standards for what is truth and right? What can possibly be used instead?

Mill's answer was utilitarianism. This term got a pretty bad rap right from the start and has never really been incorporated as such by any government as its basic guide. Americans use the concept now and then and it has often come in handy as oratorical points. It might though bear another look as one part of the design of new governance methods. With that hope in mind I will describe this a bit more fully.

The basic idea is that society should govern itself with the best options that seems to work best to produce 'happiness' and minimizes unhappiness or pain. Mill fully realized that this simple proposition was inherently contradictory, because in any large group the methods that could be used to produce more happiness for one individual could very easily reduce happiness in some other group of people. This moral dilemma is also operative, however, usually in spades with religion- based, and also for ethics- based, or other systems which have been the rule throughout history. It has been beyond difficult for us humans to design a large plan for systems which facilitate our living together in peace. Any plan that surfaces that may warrant even preliminary further

129

study usually brings out someone who right away stamps it utopian, or socialist, or you name it and that is usually that!

Democracy at this point in time, creaky and uneven though it may be, is probably still the best way even though woefully incomplete. There is much truth to Churchill when he said "Democracy is the worst form of government. Except for all the rest". Alexis de Tocqueville already in 1840 convincingly demonstrated some of the structural weaknesses inherent in a democracy. One of these is the constant risk of stupid laws and/or corrupt administrations inherent in the fickleness and short term thinking which often dominates the all- important governors, i.e., the people themselves. Nevertheless the insanely felicitous notion that we invented, that the people are the sovereigns and are the only legitimate sources and owners of any government has triumphed over everything so far and we should probably stick with the general spirit of the Declaration of Independence until we do invent a better one.

Actually of course, ideas about making governments, or life, better are not completely opaque or flabbergasting. We can almost claim as axiomatic for example, that Statelessness will practically guarantee a near maximum of unhappiness. Dictatorship is practically guaranteed also to produce no increment of happiness, and greatly increasing suffering and pain. Theocracies are not much (if any) below that level. And so on. 'Til we get to genuine Democracies, where we are at – unevenly – today. The next classification beyond that may be simply much better democracies.

In the spirit of westering, then, we might all urge our fellow Man to seek the next horizon and call on the future to do better in a clear -eyed, more rational and earthly manner. Striving for a pluralistic strong government, pluralist political institutions and pluralistic economic institutions, which were some of the gifts to the world shortly after the Glorious Revolution of 1688 and beyond, should give us a leg up at least.

Just a note here about where we are heading now,

especially since in this book we have lurched through a great variety of topics, each of which would require volumes to adequately cover. In particular, we have looked into the origin and evolution of man and his mind. We have delved into some of the science of his mind and even the philosophy surrounding his mind. I have asked – or hinted at – the question of the real nature of consciousness, and morality, even reality itself. One might expect that, at the end, this might play more significantly into some of the conclusions, recommendations and actions that are taken.

But the approach I am taking here now boils down to urge man to turn his mind to suggest methods which would produce changes in society from the exterior, i.e. rational systematic, organizational, outside, and political means. Some may say that changes must come from within the individual and then the world will change. Revolution comes from inside the individual, then to the world some say. They would agree with Thoreau, the Transcendentalist, who said "Law does not make the man free." In fact, many writers, artists, mystics, religious, antirationalists or the like might say that it is the Law which practically guarantees that man will be not only harassed by it, but chained as well. And further, that only by mentally fighting against the sources of social power, which inevitably, intrinsically and automatically inhibit free thought or actions, only then can mankind become ultimately free from exploitation by other men.

Actually, largely agreed upon enlightened law would be a quite comforting additional possession to have to keep men free. Why can't we have it both ways? We do indeed want citizens, all people, to feel within, those transcendental feelings which in any case increase individual happiness. As Harris ("Wake Up") eloquently 'proves', the tried and true techniques of exercises of the mind which create a state of mindfulness does often make one free and happy in the moment at least. Siegel too, (2017) offers numerous sensible ways from a mental health perspective to train and modify the Mind to enhance individual wellbeing. These efforts could become of great moment throughout society. What country would not cherish a comfortable majority of citizens with such advanced enlightened personal characteristics?

There ought to be ways that social organizations – especially governments – can become structured to encompass a great many of the objections that arise; this is, in fact my main aim here. It is universally agreed that there are dangers and evils in the world. It is time to lay these directly at the feet of Man and demand that better social contracts and arrangements be made by all means available. Do you believe that the people in Honduras, or Nigeria, or a hundred other failed states where the powerful piteously kill, terrify and unconscionably oppress their own people – because they can – , do you really believe that they are any different fundamentally than us in America or Norway? No, I don't either, but it is their social arrangements that probably first need changing, not the people. (I invite you to harken back to the lessons from the rat caging systems of Bruce Alexander.)

To come up with testable plans then, there may be some tools at hand, including some of the ones described earlier and below. There is a well- known human trait that is universally observed, which is a desire to be known as nice or good. Part of the psychological basis for this impulse is called reciprocal
altruism, which means that people often find it profitable to be mutually cooperative. This impulse, basically you scratch my back, I'll scratch yours is described by Wright, Greene and many others. It gives consensual moral codes their tremendous power. It is one of the most powerful of the psychological impulses that created a cohesive tribe in the first place.

Conscientiousness is another trait that works in our favor in trying to design a workable government template. Conscientiousness is one of the Big Five universal cluster of personality traits (most of which, interestingly, have a

132

heritability of about 0.5). This high heritability of most personality traits is an important finding from a great mass of careful studies. What a heritability factor of .5 means is that roughly half of the variability of a particular behavior or trait is accounted for by inherent, pretty much hardwired genetics. The other 50% is shaped by the environmental factors in that particular person's life from conception to death. One can take from that fact either (a) not much can be done with the person's basic behavior, or (b) there is still a lot of room to modify a trait. In any case, it means that personality traits have been baked into the human tribal brain to a remarkable extent. Surprisingly in fact, all of the big five personality traits, and most of the big ten personality disorder traits actually have rather high heritability, near .5 (50%). Tuschman (2013) discusses these ideas in detail vis a' vis our political enigmas.

Following this line of research it probably also means therefore, that such things as appeals to common family values, positive thinking, right thinking, love thy neighbor, strive for Tao, back to God, stand up for human rights, profound sayings, or Eastern – or western – holy books and a host of other bromides have, at best maybe a 50-50 chance of having any measurable effect. Taking any one or two of these separately, or many of them together might indeed actually be a very good thing if it could be carried off. They are all aimed at changing the mind – or maybe emotions – one person at a time and to change individuals' habits of thinking and feeling. The singularity required (one individual making the same kind of discovery each time) makes these approaches problematic. This is why Meditation, Transcendentalism and like techniques, as effective as they are for individuals, while they are to be encouraged a' la' Harris et. al. will probably not become strong enough soon enough to solve some of the pressing public needs.

The trick then is for a government – the Leviathan of Hobbes – to create and maintain those institutions and formulate those laws and regulations that strike the best

133

balance; in other words, use as the moral compass the dictum of creating the most happiness for the greatest number – or a better alternative. As will be discussed again later, it is not enough to just formulate laws, guidelines and regulations; it is imperative that a transparent, fair and impartial vast and efficient *system* of administration, oversight and accountability also be invented and put into operation. Whole new apparati to aid in monitoring and improving effectiveness of all government functions will be needed.

Slipshod government is bad government. Railing that government *is* the problem is a filmy fantasy that creates much mischief, Ronald Reagan notwithstanding. Calls for limited government, without giving details is another gauzy fiction, essentially mere lazy mumbo jumbo. It is meant by some politicians mainly to make it easier for private interests to increase power and influence. Limited government – meaning slipshod and increasingly ineffective government – does not work out well for Somalians, or Zimbabweans and a hundred other states. In fact, government is probably the most important possession most people have – for good or ill depending on where you live.

Too big government (a' la' North Korea?) is also a guarantee for increased unhappiness. While governments have to have a firm presence, they should strive to be as invisible as feasible; in other words, just right! People sometimes talk about, i.e., give lip service, to the goal of having a smart and efficient government; however, little smart effort has been actually given over to that task. Too idealistic, they say.

So, what do we do? Just give up? It turned out that Mill's best formula was pretty close to Jesus' admonition to do unto others what they would have them do unto you, or the Golden Rule. By itself, however, as abundant history tells, this is not going to turn the tide now either. Something more needs to be added. What that is, is for us, and you people of the future to decide and invent little by little. As Ivan says to his brother Alexey in Dostoevsky's "The Brothers

Karamosov" "For the secret of man's being is not only to live but to have something to live for." It should be evident by now that I think that that something might legitimately come, not from the afterlife or the supernatural, but from some of the elements within his own species and its organizations, with the aid of his science and his reason. I would again note, too, that mankind is still in its infancy, the new kid on the block and still inventing new follies of humanity. Wait until he grows up! Man, the Greatest Story Still to Tell!

Actually, utilitarianism is based on a moral principle that seems about as basic as may be possible. That is, people desire to be happy and avoid pain. Happiness is better than unhappiness. Whose happiness? Ours, and other people's too. As both Wright and Greene point out, it seems to be universally agreed that there is no limit or monopoly on happiness. Each person's happiness is just as important as anyone else's. Withholding the opportunity to reduce pain or increase another person's happiness, even by a little, would, therefore obviously be an immoral policy to pursue. Doing the opposite could lead to the objective of maximizing overall happiness.

Pinker in "The Blank Slate" provides a little guidance here.

"Rights need to be protected because when people live together their different talents and circumstances will lead some of them to possess things that others want... The goal of a peaceful and prosperous society is to minimize the use of dominance, which leads to violence and waste, and to maximize the use of reciprocity, which leads to gains that make everyone better off.... For all its limitations, human cognition is an open-ended combinatorial system, which in principle can increase its mastery over human affairs, just as it has increased its mastery of the physical and living worlds."

Joshua Greene in his "Moral Tribes" gives an excellent explanation and advocation of utilitarianism for governing use. He especially gives an extensive discussion of the Utilitarian scheme as a practical idea, including for the first time as far as I know, some theoretical and psychological bases. I will try to summarize his arguments, not necessarily to argue for their immediate adoption, but to try to encourage thinking about some process that could lead eventually to either this or a better one that might arise.

I should emphasize a central fact; regardless of what new types of governments or rules or schemes we will come up for our progress in the coming decades, the types of things that will make a person happier or unhappier then, will be about the same kinds as now because of our unchanged paleolithic brain structures. Culturally, within these parameters we will – hopefully, but not inevitably – evolve and progress. I agree wholeheartedly with the spirit in which Sam Harris ("The End of Faith") says "It is inevitable that some approaches to politics, economics, science, and even spirituality and ethics will be objectively better than their competitors (by any measure of 'better' we might wish to adopt) and gradations here will translate into very real differences in human happiness."

Greene starts from basic ideas he brings from evolutionary psychology as well as cognitive neuroscience, politics and moral philosophy. He notes the timeless tension between the individual and his own tribe and the conflicts, both psychological and moral, between tribes. Most conflicts are based on moral emotions coming from the 'automatic' brain. Each person's tribe has its own moral values and comes complete with its own norms, virtues, talking points and all the rest. So do the other tribes, only most often with different values, talking points, philosophy, vices and general view of the world. Trying to talk a different tribe into adopting another's moral code is hopeless, as all you get is assertions of rights, duties and other rationalizations of why and how their tribe's values are correct, or better, or both.

136

Another part of the problem, worldwide, is that some tribes' morality morphs into, or begins, with sin. If that tribe labels something as not only immoral, but sin, then the other tribe is absolutely helpless – – and guilty to boot. In fact, some Conservatives are prone to say that their policies are divinely mandated! Abortion and/or reproductive behavior is one current example. It is trite to say that today the two major tribes, conservative and liberal, just talk past each other. While we want basically the same things, the differing tribal worldviews lead each to endlessly repeat the same tribal philosophy and policies. Those differences, if viewed totally objectively are really quite small. And, since maybe 90% + of human behavior is driven largely subconsciously – and most of that by the emotional, automatic brain – there is a lot of inertia to be overcome by cultivating our best brains more wisely to drive human societies in a different direction. In other words, our broadened tribes can't even talk to each other, a condition we see played out every day on the News. One current example is the mantra of the conservatives when it comes to helping the poor. "Everybody should pull themselves up to reach the top." This is a silly assertion on the face of it, if they do not mean "Anybody can...." . Sure, anybody can become rich or famous, but not everybody can! In fact, ever since the dismal science of economics was invented over 200 years ago, the central idea that the world has taken from economists has been that as far as the eye can see there will always be a few rich men who have control of the tools of production and capital and a teeming multitude of the poor and low. It seems the system won't work if workers are well paid. (As far as I know, this law has never been repealed, merely mildly amended.)

There you go again!

Albert Einstein said "The most important question a person can ask himself is whether he lives in a safe or a dangerous world." Humans mostly grew up in a dangerous world, as proved by the abundant evidence of early and often violent deaths of early and historic man. However, the dangers a person encounters now is obviously different. It is also obvious though, that a vague, deep- set feeling of danger and fear is a major driver of some tribes' core values. And of course, core values can not be ripped out by the roots, no matter how ill founded some may be, as experience and science has shown. S.P. Huntington, for example describes some salient points regarding the "national identity" that has historically held various nations or tribes together: i.e., religion, language, creeds, shared culture, external threats, etc. He, among many others, believes that America is undergoing a mild (?) identity crisis and these considerations remain to be addressed. Paying close attention to the factors (Glues) that glue, or 'unglue' societies together would be useful.

Conservative tribes consider themselves, as Lakoff and others (cf. Tuschmann, 2013) have noted, to live largely in a very dangerous world. Thus, they put up the best defenses they know of to ensure their own happiness – like law and order, status quo, big army, aggressive individualism and comforting hierarchy. By definition, conservatives are primed to resist change, be authoritarian, hierarchical and wedded to looking backwards for traditional standards, stability and respect for the existing order. They deem these essential as the best defense against disasters lurking outside and inside a dicey humanity. William F. Buckley Jr. said that the role of the Conservative "is to stand athwart History and yell 'Stop'!"

It should be no surprise, but numerous studies show that the brains of Conservatives differ from liberals in many ways. One difference is that their right amygdala, the fear and disgust center, is demonstrably larger (Kanai et al 2011; Schreiber et al 2013). This squares with common political wisdom that the conservative's threshold for fear and disgust

138

is quite low. Thus, they are more likely to respond to messaging that caters to it.

Liberal tribes don't see the whole world so much of a threat, requiring only reasonable caution and prudence – and a good sized police force and army too. But mainly they desire cooperative, rational solutions to the problems that do arise. The modern and supposedly timeless and universal notion that Man needs something from above to help him, alone and helpless in a cold cruel universe is, to my mind, a whiny non-sequitur. Bonobos and wolves are equally alone in this universe – which most of us consider quite awe inspiring actually (much of the time, anway). Moreover, I think that we own this particular part of the universe and can do with it what we please. We should have enough to do without self- flagellation and self-paralysis.

"That's just it---- we don't know!" John Ashbery poem,

Utilitarianism, Values and Beyond

Greene starts the argument by stating some postulates; if there is not a universal moral truth and code, there is still no escape from moral choices. Doing nothing is not, as we could all probably heartily attest, a good option. So doing *something*, and making some tough choices about how we might proceed in the future seems the only other choice. But what should we choose, seeing that all of the previous moral codes are less than universally acceptable? One possibility is 'whatever works best'. This is basically the same as the core of utilitarianism; something like "the greatest happiness for the greatest number by whatever works best". Since Intentions are an important aspect of ethics and morality, integrating intentionality into the mix might be an imperative. Thus, *striving* to do the right thing should increase happiness, and also decrease suffering.

Remember, we have not had a new philosophy or political
139

science theory accepted on a wide scale that we can consult in at least two centuries. I should also point out something obvious: People! Surely there must be tools besides the gold standard of low political doggerel which we could choose to use to govern ourselves. Something which would rise somewhat above those that comes from our rutted tribal brains and have been tried and found wanting throughout history. The absurdity for one thing, of the creaky system by which Americans cast votes. A few hanging chads change history! A wholly electronic voting system that automatically registers bona fide voters and prompts each to cast their votes at every election, for County Sheriff to President should be easily within our reach. Surely, in another generation at least, some kind of similar system could also make it possible for everyone to easily call up the facts and circumstances regarding any issue or facet of government, even an excruciatingly narrow one, at any time at light speed. This would function, at the least, as a fact and reality checker. In fifty years or so, these types of tools should be very helpful in keeping citizens better informed and their noses to the grindstone – and oligarchs out of the paths of power!

In any event, if we follow consequential utilitarianism (what Greene calls deep pragmatism) we first find that consequences are just experiences. If we combine the idea that happiness is what matters, with the idea that we should try to maximize good consequences (experiences), we still get utilitarianism. Greene's definition of happiness is complicated and even a bit counter intuitive. At the very least, happiness is a moral value.

It's also about judging the value of all the things whose absence would diminish our happiness; like poverty, sick, alone, anxiety, prejudice and the like. Thus, between getting our happiness hot buttons pushed, and not getting those other, bad levers pulled there, is included almost everything that we value. Happiness is Greene's common currency of human values. He also says that if something doesn't affect someone's experience and doesn't contribute to either happiness or unhappiness, then it isn't a critical governance matter. Values are emotion-laden beliefs about what people ought to do to make people happier. But, in general, there is little disagreement about what does which for everybody Happiness is not, nor is utilitarianism, a utopian idea. Greene says

" Representative democracy coupled with a free press and widespread education, in short, most of the features of the current American system seems like a near ideal starting point. And the whole governmental enterprise needs to be seen as moral ,with a populace and institutions that actively support it."

Thomas Jefferson noted that democracy required an "enlightened populace. Already over time, some tribalistic cruelties like human sacrifice, witch hunts, slavery, judicial torture, inquisitions and similar have crumbled under the scrutiny of more enlightened populaces, especially as a result of application of reason to human affairs."

Pinker labels these the "forces of modernity" and are comprised of reason, science and humanism, with individual rights valued first and foremost. If these 'Big Three' are not valued and assiduously employed, bad consequences are sure to follow. And, as noted, avoiding bad consequences is often equally important as attaining good ones. (Again I would commend approaches like Sam Harris, Siegel and others to individual happiness as an ally in the fight.)

Greene emphasizes that utilitarianism is not just a cost-benefit analysis, although it would often involve these too. That is, determining happiness does not primarily relate to money. It is a moral calculation, and while it would be foolish for a government entity to not apply cost benefit analyses to projects, it is not sufficient. There is enough evidence in the literature that almost amounts to a law of human nature that there is a weak relationship between wealth – above a modest threshold – and happiness (Easterlin et al 2010). Past a certain point, wealth simply doesn't buy much happiness, notwithstanding several, mostly commercial, surveys that imply that is the case It is true that rich people express being, on average, happier than poor ones, and claim greater satisfaction also. This just reinforces one of the obvious ways to increase happiness overall. It is increases or decreases in wealth that most influences happiness due to wealth (Kahneman, 2011).

It is fairly well established by in depth studies (like Kahneman and Deaton, 2010) that, while high income improves the evaluation of their own lives, it doesn't improve emotional well-being much. People do not respond very much to monetary inducements above a certain level. Making over $75,000 per year was kind of the breakpoint where this holds true for people in the U.S. In a survey conducted by American Express they found that only one-fourth of Americans said that wealth constituted their definition of success. Most of the other familiar values, like health, congenial job, good relationships and others, scored 80%.   While absolute wealth has increased in America since the '60's, the few surveys which have compared happiness over this period showed little change. The absence of wealth, however, is a fairly sure way to reduce happiness, as well as satisfaction or any other measure of well- being. In another recent survey by a marketing agency, only 35% of people making $35,000 or less reported being very happy; near 100% of those making over $500,000 reported being happy.

Going on vacation, shared laughter, freedom from want and fear, going to a park, being able to pay all your bills; these things are all good experiences, that are of value to people, meaning happiness, or the utilitarian common currency. Just like going hungry, working for a bosshole, getting stuck daily in traffic for hours, living in cramped, stark quarters subjected to conflict all around, watching genocides on TV, etc. are bad experiences, i.e. less happiness. These ideas underlies almost all our values. Values again, are what humans, deep down – innately – desire in order to be happier. It is an unrefundable gift of inheritance from our savage ancestors. Values, however, even though they are imbedded to a considerable degree in our emotional brain, do not spring up full blown in a vacuum, but are conditioned by experience and culture.

So, both parties want what Pinker calls flourishing of humanity, whereby both parties' interests are relatively increased in a positive sum game. The search for a non-zero sum game with increased well being, he calls morality. Mutual unselfishness he says, leads to a greater flourishing of humanity, and this also comports with human nature. It also leads to a net gain of happiness or at least less major pain. (Incidentally, Egalitarianism is actually a less important goal than is commonly thought, as strict egalitarianism does not seem really well suited to human nature. In fact, biology probably makes complete egalitarianism impossible. Pure Communism too is probably dead, so that may be one less thing for politicos to worry about.)

It would seem to be obvious by now, but Governments simply must eventually come around to espousing non-zero-sum systems. Not everybody must have the same 'stuff', money, power, space and the like; but for most of those who do not, the cascading consequences from want of *sufficient* of these may come at too high a cost. Finding that sweet spot is the trick. It will be a job for iterative cycling by the philosophers, economists, political scientists and people of the future. Again, Paul Mason's and Kate Raworth's approach would seem to have much to offer in the coming struggle in the evolution of economies and governments.

143

Even in the most advanced democracies there is constant warfare between various tribes, all vying to control the all important, massive but legitimate power of the State. This could – so goes oligarchic or laissez faire thinking – enable an elite to happily funnel to themselves a plurality of the power and money benefits that flow from the efforts of All. In failed states, or even some not-failed states, this philosophy works out perfectly for those elite groups who do grab power in practice. (Think Somalia, or Haiti, Saudi Arabia, or Democratic Republic of the Congo [Zaire], Russia….. and dozens of others.) Fukuyama stated something which might be enshrined as axiomatic "when a strong State sides with a strong oligarchy, freedom faces a particularly severe threat."

It is, therefore, imperative that the good guys win the big battles for control of the institutions and apparatus of the state that sets and administers the rules. (And, of course the rule- setters must be elected by inclusive, democratic means.) Even in our pluralistic, inclusive, democratic society, there will always be the danger that a narrow set of interests will take control at the expense of the whole coalition. As an example of all this, retired Supreme Court Justice John Paul Stevens, vis 'a vis recent events felt moved to state "Money is not speech. All these decisions have done is make sure that Congress does not represent people, which it is supposed to do. Rather, Congress represents people who have the money to finance campaigns." One might also wonder, if speech is money, can speech then be Free and are millionaires more Free? (A creepy perversion of this equation was exposed in the form of a cartoon recently in the editorial page of the "Billings Gazette", where one elephant says "Corporations are People"; another says "Power to the People".)

Recent events in Egypt for example, also highlight the seemingly inevitable slide of elected governments to fall prey to special interests – in this latest case the Army;

144

thereby regressing, with drastic loss of freedom for the people and another round of increasing misery. Freedom from Fear and Freedom from Want are greatly dependent on not only individual actions but social contracts.

There are lots of objections to utilitarianism, but critics have lately proposed no congenial, concrete alternatives either. One of the objections to the theory is that you "can't measure happiness." True, there is no gold standard measure, but it is in fact being measured relatively successfully already. Ruut Veenhoven in The Netherlands is a leading force in scientific measurement and has for several years published "The World Database of Happiness". There are several academic units dedicated to its study too; for example, since 2000 the "Journal of Happiness Studies". Thus it may be expected, as there has been no theoretical reasons raised against the possibility of acceptable measures of the happiness metric, that we will have more user friendly yardsticks in the not distant future.

Daniel Gilbert finds that people in the U.S. give themselves an average Happiness Score of 75 (out of 100), which although uneven is really quite high. The most important things that make people happy ('ier) over the long term are health and satisfactory human relationships with family and friends. Ironically he says, so many people sacrifice some social relationships, and sometimes even health for money. This usually winds up reducing their happiness in the long run. He says" if a person is poor, a little money will bring a large increase in happiness, while if rich, a lot of money will bring very little more happiness."

In any case, it is not necessary to have an accurate, precise, metric; what we need is some consensual semi-quantitative tool to track groups and the direction of the overall happiness score and also its distribution. Something like a GPI, Gross Progressive Index, which is one of the ideas occasionally raised as an alternative to the GDP. GPI is just a crude suggestion, but its primary recommendation would be that it would measure, not only wealth and other positive measures of the totality of social activity, but subtracts negatives, like systemic inequality or environmental degradation.

Bhutan is the only country in the world to actually have adopted this strategy, as they use "GNH " as their official marker of progress. Gross National Happiness has been used there for several years, well under the radar. By most assessments, the whole notion has been a success and their own people are approving. They are a nation of about three-quarters of a million people, of which about a fourth live in the capital city of Thimpthu. The city has morphed into a modern clean city, with no slums and a quite happy population. They have fairly recently finalized their GNH measurement tool, which is based on a happiness scale of 1 to 10. It is somewhat like our own Census forms and administered similarly. The only results so far are from 2010 and just from a half dozen or so provinces; thus no serial measures are available yet.

Obviously, utilitarianism is not a simple one-shop solution. One cannot, one time, calculate the single Happiness Score by summing the relative happiness increases of one person times the number of people and then trying to compare that to another group who might have had their relative happiness decline because of some law or regulation. Monitoring significant shifts in the Happiness Score, however, would be worthwhile. For another thing, it might often stop being effective at the waters edge, i.e., in the relations between nations. For example, it would obviously increase the happiness, at least a little bit, if a large nation preemptively conquered a smaller nation. (How nice would it be, for us, if the U.S. would just simply try to annex Mexico – again)?

Certain it is that using something like utilitarianism should at least give the world an alternative political language. It is hopeless to leave the debates in the same old rut of each tribe trying to convince the other of the correctness of their ideology and thus the necessity for such and such a policy.

146

As noted earlier, making human organizations that are designed to allow all people to live in relative harmony requires more than sweet slogans, unctuous speeches, smooth leaders or loud argument. It will take, in fact, more of our rational brain than populations have been willing to give, at least to this point in our social evolution. The scorecard of history right up to the present time shows that precious few governments have even seriously attempted to primarily rule for the greatest good of all the people. And as Galbraith noted with alarm half a century ago, the conventional wisdom retains still today a great deal of the flavor of the economics of the nineteenth century. That is, no theoretical pathway has been advanced to show how the wages of the workers can ever be much more – if not less – than each man supporting his family at even a halfway decent level. This is true even where there is a great deal of wealth extant.

Fixing this is one of the future jobs of the economists, political scientists et. al. that I am talking about; and one that I might suggest go in on an expedited basis. Mason (2015) has already forwarded a vision worthy of attention for a new economics. Actually his concept involves a partly non-capitalistic economy in which Information, which is basically free and non-scarce, forms the basis. It has already begun, he says. Raworth (2017) also has formulated a comprehensive concept for a new monetary and economic system. She particularly decries the use of GDP as a principal measure, which measures nothing of the either happiness or health and well- being of either the people or the earth. Talking in terms of the future consequences for people's experiences, up or down, unpalatable as it may be to both sides, is likely to be in the long run interest of many more citizens of any country. For example, aborting any fetus is murder to one tribe, it is not murder and indeed can be a definite *good* to another. This, like all other future policies, requires that a *choice* be made by a government – the Leviathan – who has the authority, through

147

democratically elected representatives and other safeguards, to enforce the final decision.

It is equally incumbent on the government, and people to monitor the results of that choice. Using all the tools at hand. If people are to be the sovereign as has been the main idea underlying the whole American experiment practically since Plymouth Rock, then each citizen must bears responsibility for being more than passive vessels for receiving and accumulating blessings from their society. An unengaged, uninformed, uncaring citizen is a delinquent citizen and no degree of government perfection would long last.

Some New Governments and Tools?

Certainly we know from thousands of studies, and experience that human behavior is driven mainly by inner, emotional impulses. Which is one reason that, say, giving people all the facts about various courses of action doesn't usually alter the ultimate decision. And recall, what percentage of a person's day is run by emotions pretty much alone. Ninety percent. However, this does not mean that it is perfectly OK for governments and leaders to always appeal only to these primitive, erratic psychological instincts. Reason, facts, history, science and common sense must enter in too, or the public debates remain forever simply childish. In any case we know that the Individual brain, and also the Group brain, can over a long period be changed by culture and environment – and by thinking besides. At the present time there seems to be more appetite for the psychological warfare that feels good, than solutions to larger issues. More than one historian has observed that "reason did not seem to have a noticeably great effect on people's behavior." Modification of human social behavior, through both poetic and more mundane methods, is, however, certainly feasible and desirable.

The most obvious case of behavior modification is that of the human infant. The fetal and young infant brain is hugely

148

controlled by the 'reptilian' (mid brain) and hindbrain functioning. Obviously, parents and society have a great interest in arranging for that infant to learn all about the rules of social and personal behavior; hence the enormous arrangements for cultural evolution of their mental function for life. I cannot even attempt to assay this immense literature on child psychological development but it does point to some possible approaches to improve the process. Pinker reviews much literature for example, that indicates that young babies are much smarter and absorb more of their surroundings than we think and they indeed have minds.

We also know there is a good deal of plasticity within cultures too. While all people are fully capable of doing bad things, like murder even, we also know that most people don't, largely because of learned cultural inhibitions, as well as our innate morality circuits. Where the culture imposes fewer sanctions and there is locally ineffective or non-existent law enforcement, and even 911 calls are a useless exercise, the crime rates do go through the roof. For many of the inmates of ghettos, with nothing much really to lose, the rawest drives for sex, power and survival are naturally expressed, with more or less constant violence and dystopia therefore to be expected. Status improvement, or even survival, in such areas will almost *require* that a young man prove himself to his peers by a murder or mugging or two.

In these, essentially Stateless, areas different remedies need applying. Logically one of them would be to bring the area under control of the State. Not, hopefully, by increased jackbooted policing, but by alleviating the hopelessness and breeding grounds of the slum. (Again, it would be well to recall the rat park experiences.) There is enormous room for happiness increase here. Easier said than done, of course.

Grimly, though, studies show that already somewhat over a third of people who live in urban areas live in slums. In one slum in Calcutta 18,000 people live permanently on one acre. The percentage of slum dwellers in cities range from about

5% in the U.S. and Europe to 99% in Ethiopia, (92% in Tanzania). China's percentage is around 40%, which is typical too in South America, Asia, Middle East, and Africa. As if these data are not depressing enough, slums are rapidly expanding. (See Mike Davis' "Planet of Slums".)

And of course, psychopaths and sociopaths are a constant 3- 5% of any population, so lots of police and adequate crowbar hotels will always be needed. Add in another 5-10% with various volatile personality disorders, or brain-impaired that cannot compete without significant social assistance and it can be seen that a large residual problem remains. But all this doesn't mean that the world cannot, and should not, become a much less dangerous and unhappy place. The title of Pinker's new book, "The Better Angels of our Nature: Why Violence Has Declined" is telling. A whole lot of indicators of a better society are pointing upward worldwide, and have been since the Enlightenment. Humanity is still new at this game. Give us another century or two and then see where we are.

Recently Prof. Chris Mooney wrote "The Republican Brain: The Role of Emotion in Deciding the Fate of the Nation", in which he surveyed the psychological and neuropsychological literature in relation to political ideologies. He pointed out many clear differences in personality traits and in brain structure and function between liberal vs. conservative. Naturally, conservatives have contributed their books to this genre, one example being Ann Coulter's "If Democrats Had Any Brains They Would Be Republicans." There is getting to be quite a literature from within the political fray, so maybe some useful precepts will emerge from this political psychobiology.

Mooney in his surveys found, not at all surprisingly that almost all comparative measurements yielded clear differences (like the neuro- anatomic data on the amygdala and other brain centers noted earlier). For example, conservatives score low on the empathy scale while

150

Democrats score high. Democrats score low on the trait of hierarchy, conservatives high and so on.

Furthermore, the various parts of the brain which underlie particular reactions corresponding to general personality traits, also showed marked activity differences between the two major cultural tribes in fMRI tests. Similar results are reported by Kanai et al 2011, Schreiber et al 2013, Tuschman and others, so it will be interesting to see where this all leads over the next decades. (Not to pile on too much here, but there was recently a national Pew poll which found that 6% of U.S. scientists labeled themselves conservative Republicans. A wag commented "Science has no explanation for why this number is so high".)

Drew Westen in his book "The Political Brain" describes what a politician needs to do to get elected, complementary to Machiavelli laying out what a prince needs to do once elected/installed in power. Westen gives detailed analyses of what politicians should say, do and project to get themselves elected. His explicit directions to future political candidates are based on the inborn psychological makeup of man. He sums up the present evolved state of U.S. politics with three rules: Rule no. 1: (and the other 2 rules don't count) "Run on tapping the emotions of the electorate." Similarly, Lakoff, 2014, ("Don't Think of an Elephant") says that rule no. 1 for candidates should be "framing", i.e. use the correct language to get your agenda across. Both of these books are meant as a primer for political candidates but haven't really been road tested yet. It boils down to this: politicians, like cars or soap, are not sold by argument. 'Hysteria-mongering', for example, seems to work quite well for many a politician.

Lakoff (2002, 2008) also presented a penetrating comparison of the two major tribes vis a' vis the current U.S. culture wars scene. Their different tribes, which he calls the Strict Father model and the Nurturant Parent model are, at bottom, different worldviews of morality. Both supertribes

151

pretty much see and interpret all policy and political and social issues against their moral model, the basis of which is the Nation As (Traditional) Family. For example, reward and punishment, stability, etc., are musts for conservatives, while cooperativity and inclusiveness are musts for progressives. The world is going to the dogs according to conservatives while 'things can always be made better' from Democrats. Lakoff says "Each side sees the other as immoral, corrupt and lunkheaded." Conservatives believe that reward and punishment is the main basis of morality and generally do not believe that there are social causes of individual failures. Individuals fail or succeed, or should, by dint of moral strength by themselves. (Hence, people who don't succeed economically are not moral and pretty much deserve whatever comes their way.) Lakoff's cogent message is *how* these issues and values are framed in the political arena is critical and there is obviously much room for future evolution here.

All these behaviors, based to a great extent on deep rooted neural circuitry, need to be sorted out more rationally one of these days. Tribal warfare, while understandable and everlasting can, and must, be modified. But current tribal political positions do not spring up out of a vacuum. They are to a considerable degree learned and 'made perfect' by repetition and reward. It should be noted, maybe with some relief, that these traits do change throughout our lifespan, especially in early life and then more slowly. Unfortunately some of the changes after age 30 are towards, for example, more cynicism and authoritarianism, and less openness (Tuschmann, 2014, "Our Political Nature: The Evolutionary Origins of What Divides Us").

Similar changes in personality traits throughout an individual's lifespan have been documented recently also in our old friends the bonobos and chimps. Chimps share food readily with their fellows when they are young, but males almost completely lose this urge by the time they are young adults; in fact, fighting over food (and, like humans, many other things too) is thenceforth often fierce and bloody. Bonobos readily share food, as well as most other things throughout their lives (Hare & Kwetuenda, 2010; Wobber, Wrangham & Hare, 2010). This squares with what we saw earlier with the evolutionary differences in ape nature between the two species in so many fundamental ways, especially aggression, dominance and the like. (See also Hibbing, Alford and Smith, "Predisposed: Liberal and Conservatives and the Biology of Political Differences" which reinforces the evolutionary nature of our present situation.)

Summing up much of the above, the salient point is this: psychology and science generally is beginning to unravel the parts and function of the inner Man. The deeply etched weaknesses, foibles, biases, rages and passions of all stripes that a man contains and emanates, which were dyed into his big brain over his long prehistory, are being laid open for public scrutiny for the first time. The process is barely begun, but already some significant scientific understanding of his emotions and behaviors is beginning.

Heretofore the secret workings, the thinking and the behaviors of a man, especially those he does not want exposed to society has mainly been the province of the poet, the artist and the writer. The indiscretions and hidden crimes or weaknesses, like crassness, hate, deadly prejudice, powerlust, rage, etc. that many men have healthy doses of, seem always to surprise us. These are what Dickens, Hardy, Melville, Sinclair Lewis, Fitzgerald, Austen and a host of other artists have lived off ; i.e., the exposure of the studiously hidden arrogance or dominance, or plots, or illicit sex, and petty or mighty evil and all the rest. We shudder as we see the bone- deep meanness lurking in the hearts of Dickens' Pecksniffs or Chuzzlewits, or the smug crassness in Babbit, or the unsuspected evil secrets within a thousand other fictions. Living a lie with one ugly human secret of one kind or another is a stock in trade of fiction.

Actual exposure of shady love affairs, or criminal, corrupt or craven acts by Kings, Presidents, Ministers, or other idols of the masses is also a stock in trade of journalists and essayists and neighborhood conversations. Sometimes we shake our heads in wonder at the obtuseness, biases, cruelty, stupidity, disloyalty, or self -serving monomania, greediness, evil, or venom and a hundred other *humanisms* thus exposed. Sometimes we get a glimpse of the ugly that resides inside our own brains, but we should not be either surprised or totally depressed by it.

There are undoubtedly several millennia more of outstanding literary expositions on these foibles available for rich exploration and delight, so it is not to be deplored that science too will be teasing out some of them at the same time. The purpose, sometimes, of the literary expositions is to teach the rest of us how things are and to suggest some changes. Artists really aim at the heart of the matter – to try to find the true nature of the man called Man. The purpose of the science is often the same. Whether we are shamed into acting differently by Scrooge, or from the plain exposure by science of the biases or other innate human psychological quirks buried in our nature, the ultimate results hopefully would have some correspondence to reality and helpfulness. We should use all the tools at our disposal. Mankind should eventually demand it.

Wilson (2014) makes an intelligent and eloquent plea for the amalgamation of the humanities and science. He also lays out brilliantly, almost definitively, the scientific underpinnings of Man and his mind, his culture and his future. His synthesis and basic blueprint for man's future structure and his social survival, constitutes a significant biological achievement. I agree with him and also with an eloquent Daniel Dennett , and also, again, with the great syntheses of Gassaniga and Damasio and others, that an eventual much better, if still dim, understanding of the essence of our humanness must come from a deep appreciation of how humans evolved, first by strict

Darwinism, now by cultural and pure mentality. (Dennett refers to this as 'sorta' Darwinian evolution, or "semi-hemi-demi- Darwinism".)

Of course, the root of all evil, and maybe good too, could be money. On this score the gulf is indeed vast. As noted earlier, it should be a major focus of the new economists that will be employed by the new politics to figure out how to get the money to finance the more favorable societal good that has been decided upon. It won't do, to do things the other way around, which is what we do now. It won't do, in other words to reflexively assert we can't afford it, no matter what. One might imagine that it could eventually become hard to argue that money itself is the limiting factor in apportioning happiness or the like within our (hopefully) more functional future society. Clayton Christensen in " The Innovators Dilemma" rather convincingly says that it is no longer true that capital is scarce. Most futurists, (such as Sunstein, "Conspiracy Theories and other Dangerous Ideas") project that relative wealth will increase across the next centuries. It quite obviously has over the past several.

It is evident though, that new ways of looking at money and wealth are required. Probably at bottom, the whole conflict over money and what it means for people is imbedded in the two major tribes' differing conception of morality. Man invented money and wealth, which turn out to be just rather prosaic social and institutional, i.e. political arrangements, so he should be able to figure out better and better outcomes. The new capitalism will likely be underlain by Information – which by nature is quite free – rather than the old mantra of goods, wages, growth and hierarchical markets (e.g. Mason, Piketty, Raworth and others).

Parenthetically, it should be noted that there have actually been many rational, but faux, attempts to develop a playbook to guide a nation's politics. Propaganda has been developed to a frightening level and, of course has just been too juicy for some politicians to not put it to use. All nations use the tool to some extent. To whip up support and recruits for any war or big emergency, you see the phrase "they did not die in vain", or "yellow menace", "defend the homeland" and many other such essentially devious moral suasions meant to poison minds by paroxysms of nationalism. Hitler and his crowd are still the world champs, but the ubiquity and baseness of all the propagandist's assaults should be condemned. All of these programs are directed of course, to the basic emotional and subconscious brain wherein lies some of our enduring responses. These are often counterproductive, but also often enough the use of these cunning tools does indeed advance an individual, at the expense of the happiness of others. Politics, at present, obviously reeks to high heaven of propaganda and loose-headed rants.

 It does though, seem quite hopeless to anytime soon see a clear path through these scientific and political thickets without some additional approaches. Certainly, it should be possible with a few more decades of these kinds of studies and the inexorable rise in other related knowledge, that some useful principles and guides will be forthcoming. My main point now is: start making A cookbook of possible recipes for government. It should apply to all 200- plus countries extant (almost all basket cases, if looked at objectively). Guidelines for government doesn't seem like that is too radical a thing to start on.

  The greater problem might be to get some of the main valuable lessons into practice. A great number of people, not just politicians, haven't even quite been able to stomach Darwin yet either, adding to the difficulty. In any case, human psychology being what it is, it is well to never forget that people act mostly out of their feelings, desires and beliefs. And we must not forget the deep, genetically carved, brain modules built- in from which these spring. As Lakoff and others say, there still needs to be eventually agreed upon realities, not just those which the tribes make up from time to time.

So the point is, to try to develop some guides for new kinds of government of the future. The new political science will obviously need an upgraded toolbox and some new organization boxes, not to mention better technologies to monitor and administrate. Over the next century at least, a new political game should be evolved, including some X's and O's. It should engage the aid of science whenever appropriate as rules for the new language. There must be theory and practice. Minute, understandable, open, exact ideas and plans, please. Does anyone really know exactly what is in the tax Laws and Codes, and what kinds of mischief occurs therein? Or the Federal or State Regulations, or the regulatory rules which are supposed to regulate such things as lobbyists, trade, administration, etc. Certainly, murky government is bad government. While it may very well be true that people run on emotions most of the time, this mode does not lend itself well to matters of organization and public affairs.

Following are a few admittedly half- baked and scattered ideas meant just as general indicators of what might be possible. What we want is not revolution, but better governance through evolution. In our earlier tour of human civilizations, the picture we get of the governments that have been tried out over the past 4000 years has not been a real pretty one. Only very recently, and then only in a handful of instances, has a government aspired to, and partially succeeded at, ensuring the welfare of the people through inclusive democracies. As noted before, democracy, and surely an inclusive one seems essential.

Essential, too, is a well distributed sharing of the power and fruits of the democracy. The recent book by Acemoglu and Robinson, "Why Nations Fail: The Origins of Power, Prosperity, and Poverty" yields some excellent highway markers toward success. Powerful central, yet democratic governments which not only maintain law and order, but provide the critical institutions and incentives essential for wealth and well-being is the minimum starting point. But in order to prevent elite groups – oligarchs, or plutocrats, or both – eventually seizing a vastly disproportionate power and bending the institutionalized wealth and power machine mainly toward their own benefit, the conflicts must be between *included* groups. That the people, i.e. the vast majority must always win these ongoing conflicts – at least the consummate ones – is one of their main contentions.

Actually, the framework of new government principles is only part of the battle. In several countries, we have the basics of a great framework. We need to vastly improve the other half, too. The other half of the picture of course, is governance. By which I mean the carrying out of things i.e., the structure, function and administration of the organizations and institutions that are, or will be, set up to provide for the greatest happiness. There needs to be a vastly greater utilization of technology employed so that in all the X and O boxes of the future organizations and institutions, their policies, budget, operations and effects and effectiveness will be accessible to virtually anyone. (Is there an App for that?) How many times must disproved ideologically driven policies and practices be repeated before they are recognized by the People as failures. A major job, too, will be how to wisely harness information and put it to efficient use, and not to drown in it.

The idea is to arrange a system whereby an institution is honestly and competently run – by bureaucrats, to be sure – and doesn't, for example, become a political fiefdom with inordinate waste of time or money or worse. But more significantly, groups or politicians cannot for long hold undue power by obfuscation, made up facts, blurring of intentions or direction, propaganda and all the other common 3000 year-old methodologies. This will require some technical systems to provide, among other things, easily promulgated transparency to all and ensure uncorrupted and competent civil servants and leaders. A simple proclamation by the party in power is insufficient. There must also be acknowledgment by the people that transparency is a fact on

158

the ground. The ability by the clear majority to see what is going on, and then to rationally affect it, by vote or other means, would seem to be one corrective to the seemingly inexorable drift toward exclusionary politics and institutions. Plato saw it early on "The price of apathy is to be governed by evil men."

Future Governments simply must become more effective and efficient across the board. A government of, by and for the people, must therefore actually do things of and for the people. And not do things that are for elected or appointed leaders or their pals or camp followers that are ever present, ready, willing and able to try to bend laws, regulations, power and the like to their narrower desires. It is a hard lesson, but any worthy government answerable only to the people, must be master over any other consolidated private power block. Easy transparency is only one tool of many that must be sharpened.   Government leaders must be held to higher standards of critical thinking and performance than the ordinary 'street-think', muscle memory and tribal speak. Complex social organizations require lots of thought and care, and choices, not tribal instinctive reactions.

Just once, couldn't one of our ill-considered quick fixes work?   The Onion.

Another possible suggestion for politicos of the future is to listen, and take all children seriously! Learn the real nature of humans as a starting point for your philosophy of politics from young children and then children as they develop. That is, what do children want and need and how can you develop methods to measure effects of your governance on children? There is a great deal known, but also yet to learn about the minds of children.

Actually, a fairly well researched area is devoted to children and development of what we might call maturing morals – – but sparsely applied.   One important series of such studies involved tracking violent behaviors, from birth to maturity.

159

It was very surprising, to me, to find out that the most violent age is actually two – 2! The 'terrible twos'. Yet, maybe it shouldn't be surprising. The way we maybe should look at it, is that the world of the two year old has to be a mental tornado, with an avalanche of incomprehensible stimuli coming in. Coming with a quite undeveloped and rapidly changing brain, with no cultural background or understanding, it must be a confusing time and one to be totally experienced the only ways they know how. This would probably mean predominantly running on automatic for the most part. This means there is a whirlwind of emotions, feelings and responses, driven in good measure by the subconscious brain. This could account for the expression of the most basic of responses which involves selfishness, aggression, fear, dominance and anger as probably the major feelings. Hitting, crying, tantrums and some of the other typical behaviors might actually be predicted as the predominant social responses. Other inborn personality traits, like sympathy, altruism, sociality, cooperation etc are still nascent or submerged to a considerable degree yet.

Perhaps the closest we can come to observing feral *Homo sapiens* is on a playground. The running, the tumbling, the squealing, even the pushing and shoving, the natural joie de vivre of a species is on display. While these naturally wane throughout life, they never – naturally – disappear. Except, of course, that they – too often – do! They are caused to disappear by well- known human methods of inflicting pain and difficulty. We should try to minimize these forces by our social organizations and institutions. We should recall too, the many studies that show that personality traits can, and do change with age and life's experiences – especially traumatic, early ones. What future do we really crave for the next generations and how could we ease transitions? Why, for example, does the plan for universal Pre-K programs languish? Study of the ups or downs of children's happiness

160

as they mature to adulthood as an experiential consequence of policy seems well warranted. This seems likely to increase, in a non- zero -sum arrangement, overall happiness. And a worthy enterprise and goal for the next century.

I might propose also thinking about resurrecting a long term idea from the distant past that maybe should not be rejected out of hand, again. Thomas Jefferson over 200 years ago strongly promulgated that for maximal happiness we should strive to be a nation of yeoman farmers. Yeoman farmers were at the heart of Jefferson's fondest hopes for the continuation of the country. He, and many others since have thought that small farmers, working the soil, close to nature as being more naturally "virtuous, honest, self- reliant, free, and independent" as the best hope of the republic – or any large, sustainable government. However, in fact we were a nation of yeoman farmers for maybe only a hundred years. Even during that time, most were not particularly happy with it, as so many yeoman farmers were poor and low, struggling mightily to stay in their homes and even to feed their families. Nevertheless, the idea has much to recommend it, under certain conditions which would have to be developed.

Of course, Jefferson is also most famous for the dictum that individual liberty and pursuit of happiness was the highest goal. The most happy state he believed to lie in these independent farmers, producing wealth and greater happiness and raising their children in an increasingly happy and sustainable world. History of the U.S. since his historic – but still prescient – pronouncements has, however, shown the hegemony of yeoman farmers to have been rather short lived and withal, much less idyllic than supposed. Today, farmers are less than 3% of the population and while in some sense they truly are the backbone of the nation, yet obviously his ideal is far, far receded.

And yet, the whole fuzzy notion of a nation of virtuous free yeoman farmers having ownership of their own land, not beholden to any superior to make a living has been central to

161

the whole North American tribal mystique and psyche. "Westward the course of Empire takes it way" was sensed, and spoken by Bishop Berkeley already by 1720 and became ever after the dominant American theme. Manifest Destiny, the frontier, Eldorado, Garden of the West; whatever it was called, the pull of empty land to the west, full of limitless possibilities, with each man healthful and free is the closest thing to the essential American spirit almost from the beginning. This powerful emotional conceit has been voiced in the past three centuries in a thousand ways, by Jefferson, Whitman, Thoreau, Melville, Turner and many more. F. Scott Fitzgerald may have most poignantly captured this most unique, unparalleled and ineffably powerful- ever American legend in one of literatures's greatest passages. On the steps of Gatsby's great house overlooking the New York harbor Nick muses

"....as the moon rose higher...... gradually I became aware of the old island that flowered once for Dutch sailors' eyes.... fresh, green breast of the new world. Its vanished trees........had once pandered in whispers to the last and greatest of all human dreams; for a transitory enchanted moment man must have held his breath in the presence of this continent, compelled into an aesthetic contemplation he neither understood nor desired, face to face for the last time in history with something commensurate to his capacity for wonder."

Henry Nash Smith (1950) closely traced this central legend all the way to the end, which he declares only died, and then only partly, when the Frontier disappeared by 1890. In the end, of course, the land was not unlimited, and the yeoman farmers settled by the Homestead act in 1862 – and again in the last century – did not have limitless possibilities. Instead, too many of them lived out their lives isolated and worn out trying to 'mine' the land, ending up busted and rusted like their old hay rake. The only illimitable thing on the planet it would seem is Man's mind itself.

162

Nevertheless, the myth did not really die, and the sentiment is still strong in our DNA. There also seems to be a good deal of sound psychology in the ideal too. The dream of many young people of living a satisfactory life on their own land is powerful yet. If it could be found to be feasible in future, here's a revised, half baked plan to make it possible for more people to live on the land.

It is based partly on Jefferson's original ideas, many of which are still partly correct. That is, people living close to nature actually involved with tilling the soil and working with animals do usually derive a great deal of daily satisfaction – happiness – that does not more or less automatically accrue to people in more crowded or more difficult environments.

Wendell Berry and Louis Broomfield are among many fierce and eloquent advocates of the joys and benefits of the life on the land. One of the eloquent new young advocates of the joys and benefits of small scale farming is Ben Falk (The Resilient Farm and Homestead, 2013). He very powerfully makes the case for greater numbers of people on the land and also how to improve the land along with their health. He notes the strong mutual connection and dependence between the land and community –and, ironically that the world is full of landless people!

[As a puckish aside, a new study from Dr. Philippe Figenmann's group at the University of Geneva Hospitals showed that 'country' mice were healthier than 'city' mice. They compared a group of mice which lived in standard high quality vivariums at their laboratory in Geneva to mice living on a farm. The 'city' mice were much more susceptible to develop a range of allergies. So maybe farm kids really are healthier?]

Almost all persons (albeit an almost insignificantly small percentage nowadays) who have had occasion to live a farmers life, even as children, as I have for almost a quarter of my life, do see something of value in living closer to

163

nature. It does give a certain direction, purpose and piquancy to life. And yes, sometimes even an ineffable, supernatural-ish, 'oceanic' feeling of deep connection to the earth, if not the whole Universe and everything in it. But emphatically not when the living is precarious, penniless, and unremittingly backbreaking in competition with raging mega- agribusiness. The haunting, but very true plaint of Victor Davis Hanson "where farming is food production as part of capital exploitation" says it all.

Just a few fun facts about farms and people and land. Total U.S. farmland is about three quarter billion acres, out of 2.3 billion total land area. Roughly 50 million people (16% of the total population) live in rural areas – everything outside of metro areas. They are spread out over 70% of total land area. However, only 2.2 million families, about 6.6 million people actually are farmers and live on family farms. (There are maybe more people in prison in the U.S. than there are actual farmers.)

So by simple division, the average farm acreage today is about 420 acres. Only about half of these families however, derive their whole income from farming and collectively these produce only 20 percent of the nation's food and fiber. Very large and corporate farms, only 10% of the total farm units (~ 200,000 people) produce the other 80%. One could surely raise questions about this corporate farm model and fairly readily imagine more sustainable models.

So, if 20 percent or so of the population a century or two from now wanted to live on the land, that might become a possible goal. If say, 20 million families were able to live on the potentially 100 acres or so of current crop or grazing land, that would give each a part time job at least. Both smaller and larger acreages are also obviously feasible, but in any case the goal is not to spread people out on the land to occupy every square inch, but to increase human flourishing, one family at a time. The central goal is not to have a great many more families out vainly trying to wrest a tough living

164

as a farmer as has been so often the case. It is to enable both the land and the people to sustainably flourish and thrive. It would allow, if bare survival is more than assured by other means, (some of which are presented below) for people to gain a sense of personal contentment from the daily benefits of living in and with nature and sharing in its bounty. It would allow people *Agency*, that is, play the central role in controlling their own liv**es.** Obviously, people now residing on the land would not be evicted and replaced, but if the idea ever got traction, a better general direction – and many individual dreams – could be achieved.

[Appropo of the eviction theme, I might just note that this tactic has been tried before, and quite recently. Real tribes of the west were summarily evicted and spit out from the gears of 'progress' in service of the above great American legend. The Indians were doomed practically from the start, "when they first heard the snort of the iron horse" in the memorable dispatch of a western reporter in 1872; cf. Athearn, 1971.]

There is, though, an urgency to the central problem. Related to all of the above is the most important problem for the future. Simply put, the modern world has too many people for a unsustainable environment. The fact is, Man is outgunning the ecosystem by virtue of his unsustainable numbers and impact. I have to lend urgent support to the always unpopular idea that the human population will require some trimming over the next century. Despite near universal wrathful denunciations of this possibility, there are actually rather easy, humane and readily implementable ways this could be done without disruption or outrage. (In a word, simple encouragement of family planning.)
The urgency of the problem was highlighted recently by a clear, stiff warning about limits in a paper published in *Science* by a large number of leading scientists around the world (Steffen et al 2015). They marshaled strong evidence that mankind had already passed through four of the nine 'Boundary limits of the Earth'. The nine boundaries are

widely accepted as those marking a "safe space for the continuance of a thriving mankind". The four we have already blown through are 1) rate of key species extinctions, 2) deforestation, 3) climate change (CO_2 levels), 4) Nitrogen and Phosphorus cycle disruption.

But we do have plenty of technology and knowhow to do the job – or could have, to begin to work on all these numerous problems. Let's face it, too, numbers of industrial jobs especially, for the most part are gone forever. Leisure time and people with no urgent reason or opportunity to work full time will be the norm. Already the idea that the private sector, even including the usual government jobs can supply a good physical job potentially available for everyone, and for that job to supply a decent living is a cruel hoax. It is also obvious that by 2100, say, citizens will all have to be paid a minimum – livable, but not extravagant – dividend or salary or such, and then be free to do what they will. (Dickens has already given us maybe the perfect name for this – "Stipendiary emoluments".) It is similar to what Switzerland and India as well as many pundits and academics around the globe are considering today. Robert Reich in his recent book "Saving Capitalism" spells out how this could readily be done. Freed of the soul wrenching fear of even surviving, a great many would go into business, a great many would make extra money by doing the residual jobs, aided by high technology, that they see needs doing, and so on. (And many, as suggested above, might actually become farmers!)

What this all amounts to may simply be a reduction in the hours worked to make a living at a decent and desired level. While laziness to some degree is seen in all primitive tribes studied, and thus is a trait somewhat baked into our genes, people generally do want useful activity and work. The image of a bunch of slobs lazing around all day doing nothing and making trouble if they aren't forced to work is a popular tribal myth, but is actually a non-starter. The main job of living for most people is largely aimed at improving the trajectory of their lives for themselves, their friends and family. But they need some help from their micro and macro environments. And this is where Government and their own larger choices becomes so important. If this help would remain light, consensual, unobtrusive as possible and not massive is the goal. Perversely, it seems today that The Economy is looked at like it is some universal natural phenomenon that just exists, instead of being a wholly owned human enterprise. Humans can make, and humans can break, economies any time they would will, can't they?

  No matter what kinds of potential methods or mechanisms the best minds can come up with, there is one more major consideration that must feature prominently in the equation. Starting with Machiavelli, and especially with Hobbes, but with Locke and Bentham too, down to our own political punditry all have warned that any of the social structures that will be advanced, will also require pretty much ironclad and impartial enforcement. No more promises you have no intention of following up on. The first rule of freedom is that some of it must be curtailed for the good of all. Fortunately, we have some internal programs to control the passions that lead us to so highly prize greed, aggression, out- group hate, dominance and all the rest of a familiar litany. These need to have societal (governmental) firewalls to improve the lot – happiness – of the world of the future. Thus, enter the new Leviathan. He needs remodeling, but we will still need the powerful disinterested 'referee'. We have a good start, but what he will look like when he is finished, our great grandchildren will be interested to see – no, not interested, vitally concerned!

"Back off, man. I'm a logician!",   Blog comment, Unknown.
167

Last Words

Obviously, *how* to accomplish such monumental undertakings will be mind boggling and a good challenge for mankind over the next century or two. This is why we need you, you political scientists, economists, and all other scientists, artists, humanists and philosophers to start now.

As we all are aware, our modern times seem somehow qualitatively different now. Just within the last three quarters century, my own lifetime, the world has become a vastly different place for tribes to live in. We are all having a tough time with trying to find our place in a seemingly runaway world. But, it's not all our fault; as we saw. We were bred and raised as Nature's child, with all sorts of nasty and brutish characteristics and mostly only an automatic emotional autopilot as our guide. Not really suited to this electronic age or the future. Not well suited for large scale nation keeping either. That's why we need some new ways of thinking (and acting) about this and sorting out of social contracts.

There would seem, however, to be many reasons for hope. As noted several times earlier, there is already the beginnings of a theoretical groundwork on many fronts on which to build and progress. The task ahead may be thought of as primarily accelerated evolution of the cultural Mind. The goal is not utopia nor perfection, nor a grand philosophical solution of everything. We need not depend upon a call from the spirits out of the vasty deep. The government needs to be guided a good bit by reason and judged partly by its standards, but the people need not be continually schooled in steely Reason alone, for no reason. Many of the tools are known and available, others will need inventing.

Speaking of tools, one last unqualified suggestion if I may, towards a more near-term goal in economics. This comes from, again Galbraith, who I might summarize and paraphrase. The chief economic theory – myth – of the affluent society seems to be that full employment is the holy grail. That means that production must be at all times at full tilt, producing more and more marginally useful material goods. Full production, so goes the mantra, requires the participation in the labor market of essentially all hands. Those who don't work aren't contributing their fair share to society and are presumed to be expected to suffer the consequences; but at least there is presumed to be mostly full employment so that few starve and everybody is supposedly happy.

There is an ever increasing race to keep producing ever more stuff, much of which requires the manufacture of even the "wants" for the goods. The obsession with production and goods, which we have been conditioned for so long to accept as the normal, if not only, way to live refers, however, essentially only to the privately produced goods; that which the private sector makes, and the private sector – mostly – buys. It ignores, or belittles the public goods and services and rest of life. The worldview has been that the world is in deep poverty and therefore production, i.e. work, markets and money, becomes a necessity of deep urgency; in fact, the totally dominate task of government and society.

Galbraith concludes, however, "that we have discovered that production, in reality, is not an urgent task" and we should abandon this age- old imperative. He says balance between private and public production needs to be established. This means balance between production and labor, away from the cockeyed age-old "iron wages" hegemony. No more full employment mantras. Physical production as mentioned before, does not in fact require all hands; maybe half, due to automation, etc. It will also be possible to comfortably lessen the churning of the more trivial physical goods. Production of steel and plastic goodies by actual people as the be all and end all of peoples' lives, needs in any case to give way to more labor for human goods (like education, arts, science, environment, enjoyment, etc.). Much of the scutwork will, and should to a degree be

169

replaced (with robots and other technologies) by more creative, enjoyable and human values work. This ultimately will mean fewer hours of labor in the classical sense for survival and 'thrival'. "Economic growth", the magic words of politicians all the time, all over the world, needs to be replaced with a new ethic. Galbraith suggested over half a century ago, and now Thomas Piketty (2014), Peter Barnes (2014), Mason, Raworth and many others too, including the Pope, have joined in with some sensible methods to possibly accomplish this, with happiness increases available all around.

Thus, we want you, you people of the future, to advance, among a host of other things, the taxonomy and practice of governance. The really hard part will not be just putting down, one by one, the pretty words, and lovely theories, guidelines, or proscriptions; the hard part will be implementing it and actually creating, piece by piece, guides for working government for all. As Benjamin Franklin was reported to have said when a woman asked what they had come up with at the Constitutional Convention "A republic, Madam, if you can keep it."

Naomi Klein (2017 "No is Not Enough") very recently put forward, I think, the best program with which begin to remake governments. She lists the common 'theories' of governments and economics that a great many people more or less unconsciously hold: e.g. "that greed is good. The market rules. Money is what matters in life. The natural world is there for us to pillage. The vulnerable deserve their fate and the one percent deserve their golden towers. Anything public or commonly held is sinister and not worth protecting. We are surrounded by danger and should only look after our own. There is no alternative to any of this." She helped organize, with a large group of concerned citizens and knowledgeable experts in various fields, an on-going kind of constitutional convention to try to hash out a blueprint for the future of Canada (and the U.S.). They came up with "The Leap Manifesto"; a five page declaration of principles and programs to form the basis for a new way of governing. The 'Leap' is a clear, rather easily doable list to get the job done (at least started).

I'd like to make one thing clear about my concentrating on the Governance issue, since I will be accused of preaching enslavement of everybody by Big Brother. I am not advocating that Governments are the be- all and end-all of existence; not at all. Governments do not make people happy; but they sure as hell can make it easier or harder, or impossible. People will– should – always strive and be free to live their own <u>autonomous,</u> private lives in their own local situation, making their own personal choices about vocation, economics, friends, activities, organizations and all the rest in their own way, maximally free from interference.

. History, however, shows that to make this possible, Super-organizations and special institutions are needed to keep things from becoming all balled up and eventually almost everybody become unhappier– excepting only a lucky dominant few. Neither I, nor any major economic or political writer as far as I know, advocate for some governmental, top-down economic theory, or abandonment of vigorous free enterprise. However, not unregulated free enterprise, or predatory capitalism run amok, where money can buy politicians, governments or whatever it likes. We don't have to have everything; like McKibben, though, we would like Enough!

Make us, then, a more stable, sustainable governance system that we could use as a guide to navigate and mitigate and govern the rules of, and the effects of the ageless conflicts; that is your charge, you people of the future. Life is meant, I think, to be enjoyed. It should not be considered impractical to strive to arrange social affairs so that most people do not face unnatural, people-made barriers they must jump over or through to prove their worthiness to join the quest to enjoy the just fruits of life. Even present politicians – many of them at any rate – will admit deep down that higher planes of behavior in politics and government are not

171

only desirable but could be readily achievable, at least eventually. Large organizations – governments we're talking here – must not be primarily driven by the lowest common denominator of raw tribal myth which is impervious to normal facts or logic. Damasio writes "The time will come when the issue of human responsibility, in general moral terms as well as on matters of justice and its application, will take into account the evolving science of consciousness. Perhaps the time is now".

 We need not try to remake Man's inherent makeup, nor change everybody's philosophies; just <u>Just</u>, sensible, enforceable and buyable rules of the road of, by and for – – humans! Man, remember, is owner of the planet now, and not just a puny wannabee. He can, in significant instances, for good or evil, simply say "Be", and it "Is". It is trite, but still true that nothing on the earth is fixed, but everything, including human natural and cultural features are always changing.

   The good and bad news, though, is that these changes are not preordained and the grim possibilities always are that future changes we ourselves make, by choice or procrastination or evil, will not be happy ones.

   Finally, in ending this book I'd like to state again my main goal: it has been to provide perspective and context to the whole human saga. The main thing I wanted to say and leave with you is a little context for modern man; where he belongs in the long arc of history and prehistory and also the fact that it is all a work in progress. It boils down to a simple idea: Man is a recent arrival on the earth and his history has determined his makeup and also even his recent social accomplishments. By considering a little of the nature of his physical surroundings, and the origins of his physical body and his mental actions I hope one might gain a little different perspective on why some things are as they are – and also what one might, collectively, hope they might become in future.

We traced man out of the very first protoplasmic blob, all the way through the swarming cellular life in the seas, then through our aquatic multicellular kin of all the waters. Then through the invasion of the land and 'soon' thereafter, the appearance of our amazing primate cousins. Then very quickly – by geologic standards – came our crude caveman, the pinnacle of evolution so far. Then, starting essentially from scratch culturally with only a handful of fellows, from his first rude appearance it took him only about a million years to start the stupendous trappings of civilization. Thence only less than 12,000 years to make the modern world. Each of us now more or less equally possesses all the inherent characteristics which had been stuffed into him.

We can see clearly now: Man, that remarkable young upstart species is yet much unfinished, and has much unfinished, possibly noble, work still to do. Perhaps we should, rather than heaping some scorn on him for his last several millennia of bungling, bloody and bewildering history, just congratulate him on how much he has accomplished so far! In any case, their inheritors, our friends and neighbors, eight billion of us now inhabit the planet; thus my hope is that me and you and all of us here now alive will occasionally take stock and commit to making the future even better for a grateful posterity.

What, I wonder, will the determined hand of man, ultimately, do?

REFERENCES

Acemoglu, D. & J. A Robinson. 2012. Why Nations Fail: The Origins of Power, Prosperity, and Poverty. Crown Business, NY.

Ackerman, Diane. 2014. The Human Age: The World Shaped By Us. WW Norton, NY.

Ashbery, John. The Pie District, poem. New Yorker, June 23, 2014.

Asimov, Isaac. 1968. The Dark Ages. Houghton Mifflin, Boston.

Athearn, Robert. 1971. Union Pacific Country. Rand McNally, NY.

Babiak, P. & R. Hare. 2006. Snakes in Suits: When Psychopaths Go To Work. Harper Business, NY.

Barnes, P. 2014. With Liberty and Dividends For All: How to Save Our Middle Class When Jobs Don't Pay Enough. Berrett-Koehler Pub, San Francisco.

Bekoff, M. 2001. Social play, behaviour cooperation, fairness, trust, & the evolution of morality. Consciousness Studies 8:81-90.

Bekoff, M. & J. Pierce. 2009. Wild Justice:The Moral Lives of Animals. U. Chicago Press, Chicago.

Berry, Wendell. The Art of the Common Place. 2002. Counterpoint Presse. Washington, DC.

Berry, W. What Are People For? Counterpoint Press, Washington, DC. 2010 Ed.

Brandon, SGF, ed. 1970. Ancient Empires. Readers Digest Assoc.

Broomfield, Louis. 1945. Pleasant Valley. Harper Bros. NY.

Bushdid, C. et al. 2014. Humans can discriminate greater than one trillion olfactory stimuli.Science 343 (21 March): 1370-1372.

Carhart-Harris, R., D. Erritzoe et al. 2014. Neural correlates of the psychodelic state as determined by fMRI studies in psilocybin. Proc. Nat. Acad. Sci. 109:2138-2143.

Carrington, Richard. 1963. A Million Years of Man. World Pub. Co, Cleveland.

Chagnon, N. 1992. The Yanomamo-- The Last Days of Eden. Mariner Books.

Christensen, Clayton. 2011. The Innovator's Dilemma. Harper Business.

Churchland, P. 2011. Braintrust: What Neuroscience Tells Us about Morality. Princeton U. Press, Princeton, NJ.

Covert, M. "Scientists Model a living cell with software". Scientific American, January, 2014.

Damasio, Antonio, 2010. Self Comes to Mind:Constructing the Conscious Brain. Vintage Books.

Davis, Mike. 2006. Planet of Slums. Verso Books, NY, London.

Dawkins, Richard. 1976. The Selfish Gene. Oxford Univ. Press, NY.

de Chardin, T. 1955. The Phenomenon of Man. Harper, NY.

Decety J, Chen J., C. Harenski, K. Kiehl. 2013. An fMRI study of affective perspectives taking in individuals with psychopathy: Imagining another in pain does not evoke empathy. Frontiers in Neuroscience 7: 489.

175

Dennett,Daniel, 2017. From Bacteria to Bach and Back: The Evolution of Minds.W.W.Norton, NY.

de Tocqueville, Alexis. 1840. Democracy In America. (Numerous translated editions).

de Waal, Frans, 2013. The Bonobo and the Atheist: In Search of   Humanism Among the Primates.WW Norton, NY.

Diamond, Jared 1991/2006. The Third Chimpanzee. Harper Collins,

Dostoevsky, Fyodor. The Brothers Karamosov.

Durant, W, Durant A. 1968. The Lessons of History. Simon & Schuster, NY.

Easterlin R, A. McVey et al. 2010. The happiness-income paradox revisited.Proc. Natl. Acad. Sci. 107 (52):22463-22468.

Ehrenreich, Barbara. 2001. Nickel and Dimed: On (Not) Getting By In America. Metropolitan Books, NY.

Evrard H., Foro T., Logothetis NK. 2012. Von Economo neurons in the anterior insula of the Macaque monkey. Neuron 74:482-8.

Falk, Ben. 2013. Resilient Farm and Homestead. Chelsea Green Pub., White River, VT

Figelmann, P. et al 2017. The farming environment protects mice from allergen-induced skin contact hypersensitivity. Clin. & Exp. Allergy; DOI:10,1111/cea.12905

Folcher, M., S. Oesterle, K. Zwicky, T. Thekkottil, J. Heymoz, et al. Mind-controlled transgene expression by a wireless optogenetic designer cell implant. Nature Comm.5: 5392 doi:10 1038/nscomm6392.

Fukuyama, Francis. 2011. The Origins of Political Order: From Prehuman Times to the French Revolution. Farrar, Strauss & Giroux, NY.

Galbraith, John K. 1958. The Affluent Society. Houghton Mifflin, Boston.

Garraty, JA & P. Gay (eds). The Columbia History of the World, 1972. Harper & Row, NY.

Gazzaniga, Michael, 2018. The Consciousness Instinct: Unraveling the Mystery of How the Brain Makes the Mind.. Farrar, Straus and Giroux.

Gilbert, Daniel. 2006. Stumbling On Happiness. Knopf, NY.

Godfrey-Smith, Peter. 2016. Other Minds: The Octopus, The Sea, and the Deep Origins of Consciousness. Farrar, Straus, Giroux, NY.

Goodall, Jane. 2010. Jane Goodall: 50 Years At Gombe. Stewart, Tabori & Chang. NY.

Gould, SJ. 1999. Questioning the Millenium: A Rationalist's Guide To A Precisely Arbitrary Countdown. Harmony Books, NY.

Grabow, Rob. 2009. Voting With Our Pants Down: Why 44 Million Young Voters Have The Power To Begin The World Over Again. World Publishers Network.

Greene, J. 2103. The Moral Tribe: Emotion, Reason and The Gap between US and THEM. The Penguin Press, NY.

Griffiths R., W. Richards, U. McCann, & R. Jesse. 2006. Psilocybin can occasion mystic experiences having substantial and sustained personal meaning and spiritual significance. Psychopharmacology 187:268-183.

Haidt, Johnathan. 2013. The Righteous Mind: Why Good People Are Divided By Politics and Religion. Pantheon Books.

Hanson, Victor. 2000. The Land Was Everything.: Letters From An American Farmer. The Free Press, Simon & Schuster, NY

Hare, B. & S. Kwetuenda. 2010. Bonobos voluntarily share their own food with others. Current Biol. 20:230-231.

Hari, J. 2015. Chasing the Scream: The First and Last Days of the War on Drugs. Bloomsbury, NY.

Harris, Sam. 2004. The End of Faith: Religion, Terror, and the Future of Reason. WW Norton, NY.

Harris, Sam. 2014. Wake Up: A Guide to Spirituality Without Religion. Simon & Schuster, NY.

Hibbing JR, Alford, JR, & KB Smith. 2013. Predisposed: Liberals and Conservatives and the Biology of Political Differences. Routledge Pub.

Hohman, M. Christen, M. El-Bab, P. Buchmann and M. Fussenegger. 2014. Mind-controlled transgene expression by a wireless- powered optogenetic designer cell implant. Nat. Comm.11.

Humphrey, N. 1998. A History of the Mind: Evolution and The Birth of Consciousness. Vintage Press, London.

Huntington, SP. 2004. Who Are We: The Challenges to America's National Identity. Simon & Schuster, NY.

Jost, J., Nam, H, Amadio, & Jay van Bavel. 2014. Political Neuroscience: The Beginning of a Beautiful Friendship. Adv. In Political Psychology 35, Suppl 1.

Kaku, Michio. 2014. The Future of the Mind. Doubleday, NY.

Kahneman, D. 2011. Thinking Fast and Slow. Farrar, Straus & Giroux, NY.

Kahneman, D., Deaton A. 2010. High income improves evaluation of life but not emotional well- being. Proc. Natl. Acad. Sci. 107:16489-93.

Kanai, R., Feilden T, C Firth (yes, That Colin Firth!), and G. Rees. 2011. Political orientations are correlated with brain structures in young adults. Current Biology 21:677-680.

Kano, T. 1992. The Last Ape: Pygmy Chimpanzee Behavior and Ecology. Stanford Univ Press, Stanford, Ca.

Keysers, C. 2011. The Empathic Brain. Society of the Brain Press, London.

Klein, Naomi. 2017. No Is Not Enough: Resisting Trump's Shock Politics and Winning the World We Need. Haymarket Books, Chicago.

Koenigs, M. , M. Kruepke et al. 2012. Utilitarian moral judgment in psychopathy. Social Cognitive & Affective Neuroscience 7:708-714.

Kotler, Steven. 2015. Tomorrowland: Our Journey From Science Fiction To Science Fact. New Harvest, Amazon.

Koubeissi, M., F. Bartolomei, A. Beltagy & F. Picard. 2014. Electrical stimulation of a small brain area reversibly disrupts consciousness. Epilepsy & Behavior 37:32-35.

Krugman, P. 2005. The Great Unraveling: Losing Our Way In The New Century. W.W. Norton, NY.

Lakoff, G. 2002. Moral Politics: How Liberals and Conservatives Think. Second Ed. Univ. Chicago Press, Chicago.

Lakoff, G. 2008. The Political Mind: Why You Can't Understand 21st Century American Politics with an 18th Century Mind. Viking Books.

Lakoff, G. 2014. Don't Think of An Elephant. Know Your Values and Frame the Debate. Chelsea Green Pub., White River Junction, VT

Lewin, Richard. 1982. Thread of Life. Smithsonian Pubs., Wash.

Lorenz, K. 1949. King Solomon's Ring. Routledge.

Mason, Paul. 2015. The End of Capitalism Has Begun. The Guardian, 17 July, 2015

McKibben, Bill. 2003. Enough: Staying Human In An Engineered Age. Henry Holt & Co., NY.

Mill, JS 1874. "Nature". In: JM Robson, ed, Collected Works of John Stuart Mill. U. Toronto Press, 1969.

Miller, GA. 1981. Trends and Debates in cognitive psychology. Cognition 10: 215-226.

Montecucceo, N. F. 2006. Coherence, brain evolution, and the unity of consciousness: The evolution of planetary consciousness in the light of brain coherence research. World Future: The Journal of New Paradigm Research 62 (1): 127-133.

Mooney, C. 2012. The Republican Brain: The Science of Why They Deny Science---and Reality. John Wiley & Sons, Hoboken, NJ.

Mukamel, R, Ekstrom A, Kaplan , et al. 2010. A single - neuron response in humans during execution and observation of actions. Curr. Biol. 20:750-6.

Muller Richard. 2016 Now:The Physics of Time. WW Norton, NY.

Newton, Julie. 2007. Well-being and the Natural Environment. A Brief Overview of the Evidence. Report to the Dept. Environment, Food, & Rural Affairs, UK.

Ochsner, K., A. Bunge et al. 2002. Rethinking feelings: An fMI study of the cognitive regulation of emotion. J. Cognitive Neuroscience 14:1215-1229.

Paabo, S. 2014. Neanderthal Man: In Search of Lost Genomes. Basic Books, NY.

Pais-Viecia, M, M. Lebedev, C. Kunicki, J. Wang, M.A.L. Nicolelis. A brain-to-brain interface for realtime sharing of sensorimotor information. Scientific Reports 3(1319), 2013. 162

Pardini DA, A. Raine, K. Erickson ,R. Loeber. 2014. Lower amygdala volume in men is associated with childhood aggression, early psychopathologic traits, and future violence. Biological Psychiatry 75:73-80.

Pattee, Howard, H. & J. Raczaszek-Leonardi. 2012. Laws, Language and Life:Howard Pattee's Classic Papers On the Symbols of Physics With Contgemporary Commentary. Dordrecht, The Netherlands.

Petri, G., P. Expert, F. Turkheimer, R. Carhart-Harris, D. Nutt, P. Hellgen, & F. Vaccarino. 2014. Homological scaffold brain functional newworks. J. Royal Soc. Interface, 29 Oct. doi.org./10./1098/rsif.2014.0873.

Piketty, T. 2014. Capital in the Twenty First Century. Belknap Press/Harvard University Press.

Pinker, S. 1997. How The Mind Works. Norton, NY.

Pinker, S. 2002. The Blank Slate: The Modern Denial of Human Nature. Penguin Books, NY.

Pinker, S. 2011. The Better Angels of Our Nature: Why Violence Has Declined. Viking, Penguin Press, NY.

Raworth, Kate, 2017. Donut Economics: Seven Ways to Think Like a 21st Century Economist. Random House Business.

Reich, Robert. 2015. Saving Capitalism For The Many, Not The Few. Alfred Knopf, NY.

Rilling, JK, J. Scholz, R.Preuss, M. Glasser, B. Errangi & T. Behren. 2011. Differences between chimpanzees and bonobos in neural systems supporting social cognition. Social Cognitive & Affective Neurosciences. 7: 369-379.

Rorty, Richard. 1998. Truth and Progress: Philosophical Papers, v. 3. Cambridge Univ. Press.

Roseman L., R. Leech, A. Feilding, D. Nutt, R. Carhart-Harris. 2014. The effects of psylocybin and MDMA on between-network resting state functional connectivity in healthy volunteers. Frontiers in HumanNeuroscience. Doi.10.33899/fnhum.2014.00204.

Sapolsky, Robert. 2017. Behave: The Biology of Humans at Our Best and Worst. Penguin Press, NY.

Schreiber, D., Fonzo G, A.Simmons et al. 2013. Red brain, blue brain: Evaluation processes differ in Democrats and Republicans. PloS ONE 8(2), e52970.

Siegel, Daniel 2017. Mind: A Journey To The Heart of Being Human. W.W. Norton Co, NY..

Smith, Henry Nash. 1950. Virgin Land:The American West As Symbol And Myth. Vintage Books, NY.

Steffen, Will et al. Planetary Boundaries: Guiding human development on a changing planet. Science (online) Jan 15, 2015.

Strauss, W. & N. Howe. 1991. Generations: The History of America's Future, 1584 to 2069. Quill William Morrow, NY.

Sunstein, C. 2014. Conspiracy Theories and Other Dangerous Ideas. Simon & Schuster, NY.

Tuschmann, A. 2013. Our Political Nature: The Evolutionary Origins of What Divides Us. Prometheus Books, NY.

Veenhoven, R. 2013. World Database of Happiness.

Westen, Drew. 2007. The Political Brain: The Role of Emotion in Deciding the Fate of the Nation. Public Affairs (Perseus Book Group), NY.

Wilson, EO. 2012. The Social Conquest of Earth. Liveright Pub. NY

Wilson, EO. 2014. The Meaning of Human Existence. Liveright Publishing, NY.

Witten, IB., EE Steinberg et al 2011. Recombinase-driven rat line: Tools, techniques, and optogenetic application to dopamine- mediated reinforcement. Neuron 72:721-733.

Wittig, RM, Crockford , C. Deschner et al. 2014. Food sharing is linked to urinary oxytocin levels and bonding in related and unrelated wild chimpanzees. Proc. Roy. Soc. B, London, Jan 15, 2014.

Wobber, V., R. Wrangham, B. Hare. 2010. Bonobos exhibit delayed development of social behaviors and cognition relative to chimpanzees. Current Biology 20:226-230.

Woodward, A., J Allman. 2007. Moral intuition: Its neural substrates and normative significance. J. Physiol- Paris 101:179-202

Wright, Robert. 1994. The Moral Animal: Why We Are the Way We Are, The New Science of Evolutionary Psychology. Vintage Books, NY.

Yoosaf, K. et al. 2013. A supramolecular photosynthetic model made of a multiporphyrini array around a C60 core and a C60- imidazole derivative. Chem. Europe. J. 20:323-326.

Zimmer, C. Secrets of the Brain. National Geographic, Feb, 2014.

www.ingramcontent.com/pod-product-compliance
Lightning Source LLC
Chambersburg PA
CBHW020908180526
45163CB00007B/2663